Statistical Programs
in FORTRAN

Ronald D. Schwartz
**Associate Professor
of Business Administration
Baldwin-Wallace College**

David T. Basso
**Manager of Information
Systems, Development
Diebold, Inc.**

Reston Publishing Company, Inc.
A Prentice-Hall Company
Reston, Virginia

Library of Congress Cataloging in Publication Data

Schwartz, Ronald D.
    Statistical programs in FORTRAN.

    1. Mathematical statistics—Computer programs.
2. FORTRAN (Computer program language)   I. Basso,
David T.   II. Title.
QA276.4.S37   1983   519.5′028′5425   82-21546
ISBN 0-8359-7096-5

This textbook can be used at those institutions
which support a standard FORTRAN or WATFIV compiler.

© 1983 by Reston Publishing Company, Inc.
A Prentice-Hall Company
Reston, Virginia 22090

10 9 8 7 6 5 4 3 2 1

Printed in the United States of America

To our beloved parents, Ruth and Charles Schwartz,
and Wilma and Richard Basso

Contents

Fortran Programming Packages

Chapter 2 ☆ Analysis of Data, 15

Chapter 3 ☆ Probability, 79

Chapter 4 ☆ Probability Distributions, 96

Chapter 9 ☆ Nonparametric Tests, 223

Acknowledgments

The authors are indebted to Dr. Dieter Wassen, Chairman of the Division of Business Administration at Baldwin-Wallace College, who has been extremely supportive of this project, and to Robert Seeman of The University of Akron, whose consultation has been extremely valuable to the completion of the project. We also appreciate the assistance of Christopher Sullivan, Director of the Computer Center at Baldwin-Wallace College, and his staff, including Martin Porter, for their efforts in obtaining our program listings. Finally, to the dedicated typists and proofreaders, Pat Johnston, Donna Finnegan, Lois Fuller, and Ann Rinker—we appreciate their efforts in reading, transcribing, and organizing our "hieroglyphics" and in typing very demanding copy.

Preface

The purpose of *Statistical Programs in FORTRAN* is to provide an innovative approach to the traditional lecture method of statistics instruction.

The authors have provided a complete set of programming packages for instructing statistics through computer-supplemented instruction, making the computer an important adjunct to learning. All of the topics recommended for inclusion in the Introductory Statistics course, as set forth by the Commission on Undergraduate Programs in Mathematics (CUPM), are presented in this text.

The materials presented here are intended as an alternative to the use of "canned programming packages" which require only that the student input data and receive an answer—with no understanding of how the computer solved the problem, the logic of program flow, or the computational formulas utilized. The packages in this text are completely documented so that even an inexperienced programmer can follow the "logic flow." The student is shown the "logical reason-ing processes" of computer problemsolving. This knowledge can be used in complex applications in statistics, management science, or quantitative business analysis. These packages are especially designed for use in these courses.

This book contains a user's guide (see the Introduction) and provides the student with the necessary information for utilizing the statistical programming packages. The text is a compilation of FORTRAN statistical programming packages. Each statistical pack-age includes a statement of problem, algorithm, and program, and example problems. These packages are documented at every step to aid comprehension.

Introduction

The effectiveness of statistics instruction is maximized through the use of computer-supplemented instruction in the following ways:

- Computer programs will minimize the amount of time required for numerical computations.
- The sequence of computer operations will be more easily understood by the student through the use of algorithms and comment cards.
- Programs provide a basis for understanding the techniques of programming and provide student motivation for creating additional statistical programs.
- These packages allow students to utilize individualized learning materials independently.
- These packages give the independent researcher a tool which permits the processing of large data sets.

☆ User Information

Purpose of Materials

The purpose of these statistical packages is to supplement the lecture and/or recitation instruction of an introductory statistics course by providing the student with a powerful problemsolving tool.

Hardware Compatibility

The statistical packages can be executed on any micro-, mini-, or large computer supporting a FORTRAN compiler. These packages can be executed either in a batch or an interactive mode.

Use of Statistical Packages

Students are to utilize statistical packages only after they have received instruction on the statistical concept studied. Thus the package acts as a "reinforced learning technique."

Programming Knowledge Needed

Students need to be given seven one-hour programming sessions using the FORTRAN language. The topics should include:

1. Introduction to computer problemsolving
2. Elements of FORTRAN
3. Arithmetic operations
4. Input/output
5. Control statements
6. Looping and the DO statement
7. Arrays and subscripted variables.

Fortran Programming Packages

Chapter 1
Summation Notation

A. Statement of Problem

Develop a computer program which will sum a set of *N* variates: X_1, $X_2, \ldots, X_N$, as given in the equation below:

$$\sum_{i=1}^{N} X_i = X_1 + X_2 + \ldots + X_N$$

B. Algorithm

1. Determine the number (N) of data points (X) in the series to be used in the analysis.
2. Insure that the number (N) of data points (X) is 1 or more.
3. Accumulate the sum of the data points (SUMX) for the entire series of *N* numbers.

3

C. General Program

```
C
C
C     ***********************************************************************
C     *  THIS PROGRAM PERFORMS A SUMMATION OF A SERIES OF DATA POINTS.      *
C     ***********************************************************************
C
C
C
C     ***********************************************************************
C     *  STEP 1.  DETERMINE THE NUMBER (N) OF DATA POINTS (X) IN THE        *
C     *  SERIES TO BE USED IN THE ANALYSIS.                                 *
C     ***********************************************************************
C
      READ (5, 500) N
  500 FORMAT (I6)
C
C
C     ***********************************************************************
C     *  STEP 2.  TEST THE NUMBER (N) OF DATA POINTS TO INSURE IT IS 1 OR*
C     *  MORE.  IF IT IS NOT, THE PROGRAM STOPS.                            *
C     ***********************************************************************
C
      IF (N .LT. 1) STOP
C
C
C     ***********************************************************************
C     *  STEP 3.  ACCUMULATE THE SUM OF ALL THE DATA POINTS (SUMX).         *
C     ***********************************************************************
C
C
C
C     ***********************************************************************
C     *  INITIALIZE THE ACCUMULATOR OF THE SUM OF ALL DATA POINTS (SUMX) *
C     *  BY SETTING IT EQUAL TO ZERO.                                       *
C     ***********************************************************************
C
      SUMX = 0.0
C
C     ***********************************************************************
C     *  INITIALIZE THE COUNTER (J) BY SETTING IT EQUAL TO ZERO;  THIS      *
C     *  WILL BE USED TO COUNT THE NUMBER OF DATA POINTS (X) THAT HAVE      *
C     *  BEEN PROCESSED.                                                    *
C     ***********************************************************************
C
      J = 0
C
C     ***********************************************************************
C     *  GET THE NEXT DATA POINT (X).                                       *
C     ***********************************************************************
C
   20 READ (5, 501) X
  501 FORMAT (F10.3)
C
C     ***********************************************************************
C     *  ACCUMULATE THE SUM OF THE DATA POINTS (SUMX) BY ADDING THE VALUE*
C     *  OF THE CURRENT DATA POINT (X) TO THE PREVIOUS SUM OF THE DATA      *
C     *  POINTS (SUMX).                                                     *
C     ***********************************************************************
C
      SUMX = SUMX + X
C
C     ***********************************************************************
C     *  ADD ONE TO THE COUNTER (J) TO INDICATE THAT ANOTHER DATA POINT     *
C     *  (X) HAS BEEN PROCESSED.                                            *
C     ***********************************************************************
C
      J = J + 1
```

```
C
C      **************************************************************
C      * IF ALL DATA POINTS (X) HAVE NOT BEEN PROCESSED. BRANCH BACK TO  *
C      * GET ANOTHER;  OTHERWISE, CONTINUE TO THE NEXT INSTRUCTION.      *
C      **************************************************************
C
       IF (J .NE. N) GO TO 20
C
C      **************************************************************
C      * PRINT OUT THE SUM OF THE DATA POINTS (SUMX) IN THIS SERIES      *
C      **************************************************************
C
       WRITE (6, 600) SUMX
  600  FORMAT (1X,30HTHE SUM OF THE DATA POINTS IS , F10.3)
C
C      **************************************************************
C      * STOP THE PROGRAM.                                              *
C      **************************************************************
C
       STOP
C
C      **************************************************************
C      * INDICATE THE END OF THE SOURCE PROGRAM.                        *
C      **************************************************************
C
       END
```

D. Example Problems

If $X_1 = 2$, $X_2 = 5$, $X_3 = 8$, and $X_4 = 12$, find

$$\sum_{i=1}^{4} X_i$$

Example 1

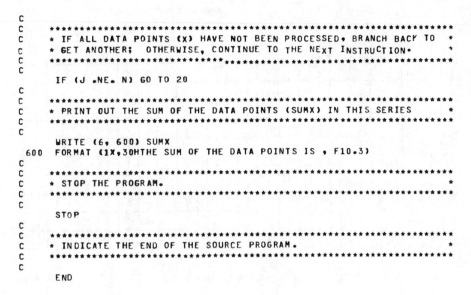

Input Data

THE SUM OF THE DATA POINTS IS 27.000

Solution

During a twenty-four-hour period, ten individual samplings of a production line showed the following number of defective parts: 22, 18, 40, 16, 12, 17, 23, 41, 29, 33. Find the total number of defective parts within this twenty-four-hour period.

Example 2

Input Data

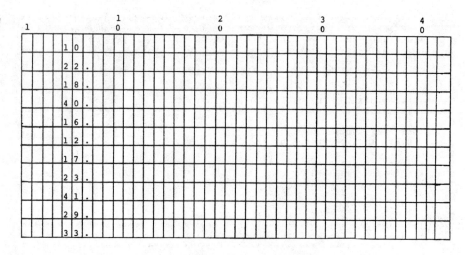

Solution   THE SUM OF THE DATA POINTS IS 251.000

☆ Summation of *N* Variates Using Subscripted Variables

A. Statement of Problem

Develop a computer program which will sum a set of N variates: X_1, $X_2, \ldots, X_N$, as given in the equation below using subscripted variables:

$$\sum_{i=1}^{N} X_i = X_1 + X_2 + \ldots + X_N$$

B. Algorithm

1. Determine the number (N) of data points (X) in the series to be used in the analysis.
2. Insure that the number (N) of data points (X) is 1 or more.
3. Read the data points (X) into an array (XARRAY).
4. Accumulate the sum of all the data points (SUMX).

C. General Program

```
C
C
C     *******************************************************************
C     * THIS PROGRAM PERFORMS A SUMMATION OF A SERIES OF DATA POINTS US-*
C     * ING SUBSCRIPTS.                                                 *
C     *******************************************************************
C
C
C
C     *******************************************************************
C     * INFORM THE COMPUTER THAT XARRAY WILL BE AN ARRAY OF DATA POINTS *
C     * WITH A MAXIMUM OF 300 ELEMENTS.  THIS ARRAY, OR VECTOR, WILL BE *
```

```
C     * USED TO STORE THE DATA POINTS (X) IN THE COMPUTER'S MEMORY.      *
C     * NOTE THAT THE CHOICE OF 300 AS A LIMIT IS ARBITRARY.            *
C     *****************************************************************
C
      DIMENSION XARRAY(300)
C
C     *****************************************************************
C     * STEP 1.  DETERMINE THE NUMBER (N) OF DATA POINTS (X) IN THE SER-*
C     * IES TO BE USED IN THE ANALYSIS.                                *
C     *****************************************************************
C
      READ (5, 500) N
  500 FORMAT (I6)
C
C     *****************************************************************
C     * STEP 2.  TEST THE NUMBER (N) OF DATA POINTS (X) TO INSURE IT IS *
C     * 1 OR MORE.  IF IT IS NOT, THE PROGRAM STOPS.                   *
C     *****************************************************************
C
      IF (N .LT. 1) STOP
C
C     *****************************************************************
C     * TEST THE NUMBER (N) OF DATA POINTS (X) IN THE SERIES TO INSURE  *
C     * IT IS 300 OR LESS.  IF IT IS NOT, THE PROGRAM STOPS.  NOTE THAT *
C     * THE CHOICE OF 300 IS AN ARBITRARY RESTRICTION OF THE PROGRAM.   *
C     * HOWEVER, THIS RESTRICTION MUST BE ENFORCED SINCE THE ARRAY      *
C     * XARRAY CAN HOLD A MAXIMUM OF 300 ELEMENTS.                      *
C     *****************************************************************
C
      IF (N .GT. 300) STOP
C
C     *****************************************************************
C     * STEP 3.  READ ALL THE DATA POINTS (X) INTO AN ARRAY (XARRAY).   *
C     *****************************************************************
C
C
C     *****************************************************************
C     * INITIALIZE THE COUNTER AND SUBSCRIPT (J) BY SETTING IT EQUAL TO *
C     * ONE.  THIS WILL BE USED TO INDICATE THE NUMBER OF THE DATA POINT*
C     * (X) ABOUT TO BE READ; IT WILL ALSO BE USED AS A SUBSCRIPT TO    *
C     * REFERENCE ANY PARTICULAR ELEMENT OF THE ARRAY.                 *
C     *****************************************************************
C
      J = 1
C
C     *****************************************************************
C     * GET THE NEXT DATA POINT (X).                                   *
C     *****************************************************************
C
   20 READ (5, 501) X
  501 FORMAT (F10.3)
C
C     *****************************************************************
C     * AT THIS POINT, J INDICATES THE NUMBER OF THE DATA POINT JUST    *
C     * READ.  THE JTH DATA POINT WILL NOW BE STORED IN THE JTH ELEMENT *
C     * OF THE ARRAY CALLED XARRAY.  NOTE THAT XARRAY(J) INDICATES THE  *
C     * JTH ELEMENT OF THE ARRAY CALLED XARRAY.                        *
C     *****************************************************************
C
      XARRAY (J) = X
C
C     *****************************************************************
C     * ADD ONE TO THE COUNTER AND SUBSCRIPT (J) TO INDICATE THE NUMBER *
C     * OF THE NEXT DATA POINT ABOUT TO BE READ AND THE NEXT ELEMENT OF *
C     * THE ARRAY ABOUT TO BE USED.                                    *
```

```
C     ***************************************************************
C
C         J = J + 1
C
C     ***************************************************************
C     * IF ALL THE DATA POINTS (X) HAVE NOT BEEN PROCESSED, BRANCH BACK *
C     * TO GET ANOTHER; OTHERWISE, CONTINUE TO THE NEXT INSTRUCTION.    *
C     ***************************************************************
C
C         IF (J .LE. N) GO TO 20
C
C     ***************************************************************
C     * STEP 4.  ACCUMULATE THE SUM OF ALL THE DATA POINTS.         *
C     ***************************************************************
C
C
C     ***************************************************************
C     * INITIALIZE THE ACCUMULATOR OF THE SUM OF ALL DATA POINTS (SUMX) *
C     * BY SETTING IT EQUAL TO ZERO.                                *
C     ***************************************************************
C
C         SUMX = 0.0
C
C     ***************************************************************
C     * RE-INITIALIZE THE COUNTER AND SUBSCRIPT (J) BY SETTING IT EQUAL *
C     * TO ONE.  THIS WILL AGAIN BE USED TO COUNT THE DATA POINTS AS    *
C     * THEY ARE PROCESSED.  IT WILL ALSO BE USED AS A SUBSCRIPT TO REF-*
C     * ERENCE ANY PARTICULAR ELEMENT OF THE ARRAY.                 *
C     ***************************************************************
C
C         J = 1
C
C     ***************************************************************
C     * ACCUMULATE THE SUM OF ALL THE DATA POINTS BY ADDING THE VALUE OF*
C     * THE JTH DATA POINT (XARRAY(J)) TO THE PREVIOUS VALUE OF THE SUM *
C     * OF THE DATA POINTS (SUMX).                                  *
C     ***************************************************************
C
   30     SUMX = SUMX + XARRAY(J)
C
C     ***************************************************************
C     * ADD ONE TO THE COUNTER AND SUBSCRIPT (J) TO INDICATE THE NUMBER *
C     * OF THE NEXT DATA POINT TO BE PROCESSED AND THE NEXT ELEMENT OF  *
C     * THE ARRAY TO BE REFERENCED.                                 *
C     ***************************************************************
C
C         J = J + 1
C
C     ***************************************************************
C     * IF ALL THE DATA POINTS HAVE NOT BEEN PROCESSED, BRANCH BACK TO  *
C     * GET ANOTHER.  OTHERWISE CONTINUE TO THE NEXT INSTRUCTION.   *
C     ***************************************************************
C
C         IF (J .LE. N) GO TO 30
C
C     ***************************************************************
C     * PRINT OUT THE SUM OF THE DATA POINTS (SUMX) IN THIS SERIES. *
C     ***************************************************************
C
C         WRITE (6, 600) SUMX
  600     FORMAT (1X, 30HTHE SUM OF THE DATA POINTS IS , F10.3)
C
C     ***************************************************************
C     * STOP THE PROGRAM.                                          *
```

```
C     ****************************************************************
C
      STOP
C
C     ****************************************************************
C     *  INDICATE THE END OF THE SOURCE PROGRAM.                     *
C     ****************************************************************
C
      END
```

D. Example Problems

The number of rakes received by a wholesale distributor for a ten-day period were 1120, 1150, 2110, 3130, 2115, 2109, 1735, 1804, 1951, and 2800. Find the total number of rakes received during this time period.

Example 1

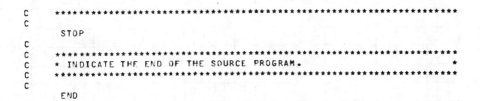

Input Data

THE SUM OF THE DATA POINTS IS 20024.000

Solution

A computer center received the following number of jobs during an eight-hour period: 30, 20, 17, 21, 10, 13, 19, 28. Find the total number of jobs received during this eight-hour period.

Example 2

Input Data

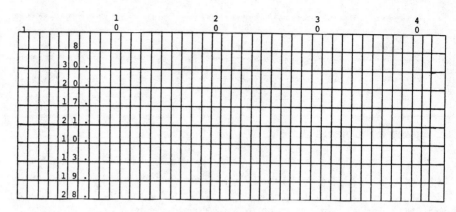

Solution   THE SUM OF THE DATA POINTS IS 158.000

☆ Summation of the Squared Values of *N* Data Points

A. Statement of Problem

Given a set of N variates, $X_1, X_2, \ldots, X_N$, develop a computer program which will calculate the sum of the squared values of the N data points as indicated in the equation below:

$$\sum_{i=1}^{N} X_i^2 = X_1^2 + X_2^2 + \ldots + X_N^2$$

B. Algorithm

1. Determine the number (N) of data points (X) in the series to be used in the analysis.

2. Insure that the number (N) of data points (X) is 1 or more.

3. Accumulate the sum of the squares of the data points (SUMX2) for the entire series of N numbers.

   a. Calculate the squared value (XSQRED) of a data point (X) by multiplying that data point by itself.

   b. Accumulate the sum of the squared values (XSQRED).

C. General Program

```
C
C
C     ****************************************************************
C     * THIS PROGRAM PERFORMS A SUMMATION OF THE SQUARED VALUES OF A  *
C     * SERIES OF DATA POINTS.                                        *
C     ****************************************************************
C
```

```
C
C     ***********************************************************************
C     * STEP 1.  DETERMINE THE NUMBER (N) OF DATA POINTS (X) IN THE       *
C     * SERIES TO BE USED IN THE ANALYSIS.                               *
C     ***********************************************************************
C
      READ (5, 500) N
  500 FORMAT (I6)
C
C     ***********************************************************************
C     * STEP 2.  TEST THE NUMBER (N) OF DATA POINTS TO INSURE IT IS 1 OR *
C     * MORE.  IF IT IS NOT, THE PROGRAM STOPS.                          *
C     ***********************************************************************
C
      IF (N .LT. 1) STOP
C
C     ***********************************************************************
C     * STEP 3.  ACCUMULATE THE SUM OF THE SQUARES OF ALL THE DATA       *
C     * POINTS (SUMX2).                                                  *
C     ***********************************************************************
C
C
C     ***********************************************************************
C     * INITIALIZE THE ACCUMULATOR OF THE SUM OF THE SQUARES OF ALL DATA *
C     * POINTS (SUMX2) BY SETTING IT EQUAL TO ZERO.                      *
C     ***********************************************************************
C
      SUMX2 = 0.0
C
C     ***********************************************************************
C     * INITIALIZE THE COUNTER (J) BY SETTING IT EQUAL TO ZERO;  THIS    *
C     * WILL BE USED TO COUNT THE NUMBER OF DATA POINTS (X) THAT HAVE    *
C     * BEEN PROCESSED.                                                  *
C     ***********************************************************************
C
      J = 0
C
C     ***********************************************************************
C     * GET THE NEXT DATA POINT (X).                                     *
C     ***********************************************************************
C
   20 READ (5, 501) X
  501 FORMAT (F10.3)
C
C     ***********************************************************************
C     * STEP 3(A).  CALCULATE THE SQUARED VALUE (XSQRED) OF THE DATA     *
C     * POINT (X).                                                       *
C     ***********************************************************************
C
      XSQRED = X * X
C
C     ***********************************************************************
C     * STEP 3(B).  ACCUMULATE THE SUM OF THE SQUARED VALUES (XSQRED).   *
C     ***********************************************************************
C
      SUMX2 = SUMX2 + XSQRED
C
C     ***********************************************************************
C     * ADD ONE TO THE COUNTER (J) TO INDICATE THAT ANOTHER DATA POINT   *
C     * (X) HAS BEEN PROCESSED.                                          *
C     ***********************************************************************
C
      J = J + 1
C
C     ***********************************************************************
C     * IF ALL DATA POINTS (X) HAVE NOT BEEN PROCESSED, BRANCH BACK TO   *
```

```
C     * GET ANOTHER;  OTHERWISE, CONTINUE TO THE NEXT INSTRUCTION.       *
C     ***********************************-*********************************
C
      IF (J .NE. N) GO TO 20
C
C     ****************************************************************
C     * PRINT OUT THE SUM OF THE SQUARES OF THE DATA POINTS (SUMX2) IN  *
C     * THIS SERIES.                                                    *
C     ****************************************************************
C
      WRITE (6, 600) SUMX2
  600 FORMAT (1X,40HTHE SUM OF THE SQUARES OF THE VALUES IS , F10.3)
C
C     ****************************************************************
C     * STOP THE PROGRAM.                                              *
C     ****************************************************************
C
      STOP
C
C     ****************************************************************
C     * INDICATE THE END OF THE SOURCE PROGRAM.                        *
C     ****************************************************************
C
      END
```

D. Example Problems

Example 1   Given the variates $X_1 = 1$, $X_2 = 3$, $X_3 = 9$, and $X_4 = 11$, find

$$\sum_{i=1}^{4} X_i^2$$

Input Data

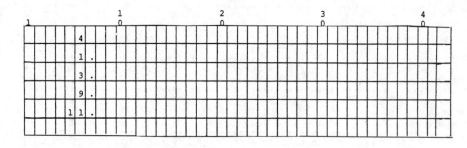

Solution   THE SUM OF THE SQUARES OF THE VALUES IS 212.000

Example 2   Find the sum of the squares of the first ten integers.

Input Data

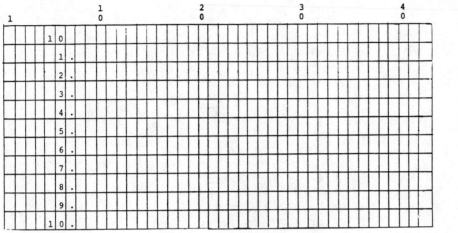

THE SUM OF THE SQUARES OF THE VALUES IS 385.000

Solution

☆ **Exercises**

1. Given that $X_1 = 3$
    $$X_2 = 4$$
    $$X_3 = 6$$
    $$X_4 = 8$$
    $$X_5 = 12$$

    find: (a) $\displaystyle\sum_{i=1}^{5} X_i$

    (b) $\displaystyle\sum_{i=2}^{5} X_i^2$

2. The number of defective parts produced at a particular plant during an eight-hour period are listed below:

    2   4   6   8   10   12   8   6   4   12   8   10   14
    10   6   9   8   4   10   2   6   3   7   8   12   10
    16   10   4   3   2   1   9

3. Find the total number of defective parts produced at this plant during this eight-hour period.

    Given $X_1 = 3$ $Y_1 = 12$
    $$X_2 = 4 \qquad\qquad Y_2 = 6$$
    $$X_3 = 6 \qquad\qquad Y_3 = 9$$
    $$X_4 = 8 \qquad\qquad Y_4 = 12$$
    $$X_5 = 12 \qquad\qquad Y_5 = 14$$

    write a computer program to compute:

    $$\sum_{i=1}^{5} X_i Y_i$$

Chapter 2
Analysis of Data

☆ **Calculation of the Arithmetic Mean for a Sequence of *N* Numbers**

A. Statement of Problem

Given a set of *N* variates X_1, X_2,..., X_N, and using the calculating formula:

$$m = \frac{\sum\limits_{i=1}^{N} X_i}{N} = \frac{X_1 + X_2 + ... + X_N}{N}$$

write a computer program to find the arithmetic mean of *N* variates.

B. Algorithm

1. Determine the number (N) of data points (X) in the series for which the arithmetic mean is desired.
2. Insure that the number (N) of data points (X) to be used is 1 or more.
3. Calculate the arithmetic mean (AVG) of the series.

15

a. Accumulate the sum of the data points (SUM) for the entire series of N numbers.

b. Calculate the arithmetic mean (AVG) by dividing the sum of data points (SUM) by the number (N) of data points in the series.

C. General Program

```
C
C
C    *********************************************************************
C    * THIS PROGRAM CALCULATES THE ARITHMETIC MEAN OF A SERIES OF DATA *
C    * POINTS.                                                         *
C    *********************************************************************
C
C
C
C    *********************************************************************
C    * STEP 1.  DETERMINE THE NUMBER (N) OF DATA POINTS (X) IN THE SER-*
C    * IES FOR WHICH THE ARITHMETIC MEAN IS DESIRED.                   *
C    *********************************************************************
C
      READ (5, 500) N
  500 FORMAT (I6)
C
C
C    *********************************************************************
C    * STEP 2.  TEST THE NUMBER (N) OF DATA POINTS (X) TO INSURE THAT  *
C    * IT IS 1 OR MORE.  IF IT IS NOT, THE PROGRAM STOPS.              *
C    *********************************************************************
C
      IF (N .LT. 1) STOP
C
C
C    *********************************************************************
C    * STEP 3.  CALCULATE THE ARITHMETIC MEAN (AVG) OF THE SERIES.     *
C    *********************************************************************
C
C
C
C    *********************************************************************
C    * STEP 3(A).  ACCUMULATE THE SUM OF ALL THE DATA POINTS (SUM) IN  *
C    * THE SERIES.                                                     *
C    *********************************************************************
C
C
C
C    *********************************************************************
C    * INITIALIZE THE ACCUMULATOR OF THE SUM OF ALL THE DATA POINTS    *
C    * (SUM) BY SETTING IT EQUAL TO ZERO.                             *
C    *********************************************************************
C
      SUM = 0.0
C
C
C    *********************************************************************
C    * INITIALIZE THE COUNTER (K) BY SETTING IT EQUAL TO ZERO; THIS    *
C    * WILL BE USED TO COUNT THE NUMBER OF DATA POINTS (X) THAT HAVE   *
C    * BEEN PROCESSED.                                                 *
C    *********************************************************************
C
      K = 0
C
C
C    *********************************************************************
C    * GET THE NEXT DATA POINT.                                        *
C    *********************************************************************
C
    5 READ (5, 501) X
  501 FORMAT (F10.3)
```

```
C
C
C     ****************************************************************
C     * ACCUMULATE THE SUM OF ALL THE DATA POINTS (SUM) BY ADDING THE *
C     * VALUE OF THE CURRENT DATA POINT (X) TO THE PREVIOUS SUM OF ALL  *
C     * THE DATA POINTS (SUM).                                        *
C     ****************************************************************
C
      SUM = SUM + X
C
C     ****************************************************************
C     * ADD 1 TO THE COUNTER (K) TO INDICATE THAT ANOTHER DATA POINT (X)*
C     * HAS BEEN PROCESSED.                                          *
C     ****************************************************************
C
      K = K + 1
C
C     ****************************************************************
C     * IF ALL THE DATA POINTS (X) HAVE NOT YET BEEN PROCESSED, BRANCH *
C     * BACK TO GET ANOTHER; OTHERWISE, CONTINUE TO THE NEXT INSTRUC- *
C     * TION.                                                        *
C     ****************************************************************
C
      IF (K .NE. N) GO TO 5
C
C     ****************************************************************
C     * STEP 3(B).  CALCULATE THE ARITHMETIC MEAN (AVG) BY DIVIDING THE *
C     * SUM OF ALL THE DATA POINTS (SUM) BY THE NUMBER (N) OF DATA    *
C     * POINTS IN THE SERIES.                                        *
C     ****************************************************************
C
      AVG = SUM / N
C
C     ****************************************************************
C     * PRINT OUT THE ARITHMETIC MEAN (AVG) OF THE SERIES OF DATA     *
C     * POINTS.                                                      *
C     ****************************************************************
C
      WRITE (6, 600) AVG
  600 FORMAT (1X,37HTHE ARITHMETIC MEAN OF THE SERIES IS , F10.3)
C
C     ****************************************************************
C     * STOP THE PROGRAM.                                            *
C     ****************************************************************
C
      STOP
C
C     ****************************************************************
C     * INDICATE THE END OF THE SOURCE PROGRAM.                      *
C     ****************************************************************
C
      END
```

D. Example Problems

Find the arithmetic mean for the following set of variates: 2, 3, 5, 7,   *Example 1*
and 9.

Input Data

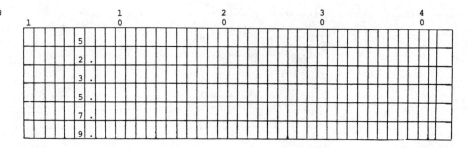

Solution   THE ARITHMETIC MEAN OF THE SERIES IS 5.200

Example 2   A production manager of a tire manufacturing firm wants to know the average number of passenger tires produced on a certain day by each of its eight plants. The number of passenger tires produced in the eight plants on a particular day were 1200, 1105, 1002, 1321, 1286, 1115, 1342, and 1368. Find the average number of passenger tires produced on that day.

Input Data

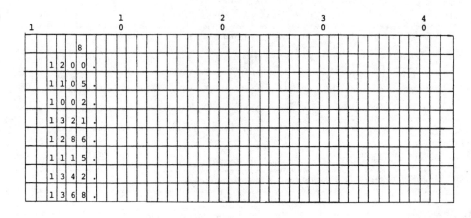

Solution   THE ARITHMETIC MEAN OF THE SERIES IS 1217.375

☆ Calculation of a Median for a Series of *N* Data Points

A. Statement of Problem

Given a set of N numbers, X_1, X_2,..., X_N written in either ascending or descending numerical order, use the definitional formulas for computing a median as defined as:

$$M_d = X_{\left(\frac{N+1}{2}\right)} \qquad \text{if } N \text{ is odd}$$

and as:

$$M_d = \frac{X_{\frac{N}{2}} + X_{\left(\frac{N}{2}+1\right)}}{2} \qquad \text{if } N \text{ is even}$$

Write a computer program to calculate the median for a sequence of *N* numbers.

B. Algorithm

1. Assume a series of data points (X) that is in either ascending or descending numerical order.

2. Determine the number (N) of data points (X) in the series.

3. If the number (N) of data points (X) is odd, the median (XMED) is the [(N + 1)/2]th data point.

4. If the number (N) of data points (X) is even, the median (XMED) is the sum of (N/2)th and [(N/2) + 1] data points divided by 2.

C. General Program

```
C
C
C     ********************************************************************
C     *  THIS PROGRAM DETERMINES THE MEDIAN OF A SERIES OF DATA POINTS.  *
C     ********************************************************************
C
C
C
C     ********************************************************************
C     *  INFORM THE COMPUTER THAT X WILL BE A VECTOR OF DATA POINTS WITH *
C     *  A MAXIMUM OF 300 ELEMENTS.  THIS VECTOR WILL BE USED TO STORE   *
C     *  THE DATA POINTS IN THE SERIES FOR FUTURE REFERENCE.  NOTE THAT  *
C     *  THE CHOICE OF 300 AS A LIMIT IS ARBITRARY.                      *
C     ********************************************************************
C
      DIMENSION X(300)
C
C     ********************************************************************
C     *  STEP 1.  NOTE THAT THE PROGRAM ASSUMES THAT THE DATA POINTS ARE *
C     *  IN EITHER ASCENDING OR DESCENDING NUMERICAL ORDER.              *
C     ********************************************************************
C
C
C     ********************************************************************
C     *  STEP 2.  DETERMINE THE NUMBER (N) OF DATA POINTS IN THE SERIES. *
C     *  THIS WILL BE ACCOMPLISHED BY COUNTING THE DATA POINTS AS THEY   *
C     *  ARE READ.  IT WILL BE ASSUMED THAT A DATA POINT WITH A VALUE OF *
C     *  -98765 WILL SIGNIFY THE END OF THE SERIES;  NOTE, HOWEVER, THAT *
C     *  THE DATA POINT WITH THIS VALUE IS NOT CONSIDERED PART OF THE    *
C     *  SERIES.                                                         *
C     ********************************************************************
C
C
C     ********************************************************************
C     *  INITIALIZE THE COUNTER AND SUBSCRIPT (N) BY SETTING IT EQUAL TO *
C     *  ONE.  THIS WILL BE USED TO INDICATE THE NUMBER OF THE DATA POINT*
```

```
C     * (X) ABOUT TO BE PROCESSED; IT WILL ALSO BE USED AS A SUBSCRIPT  *
C     * TO REFERENCE ANY PARTICULAR DATA POINT IN THE VECTOR.           *
C     *****************************************************************
C
      N = 1
C
C     *****************************************************************
C     * GET THE NEXT DATA POINT AND STORE IT TEMPORARILY IN THE VARIABLE*
C     * CALLED TEMPX.                                                   *
C     *****************************************************************
C
 10   READ (5, 500) TEMPX
 500  FORMAT (F10.3)
C
C     *****************************************************************
C     * CHECK THE VALUE OF THE TEMPORARY STORAGE VARIABLE (TEMPX) TO DE-*
C     * TERMINE IF IT IS THE SPECIAL VALUE -98765.  IF IT IS, BRANCH TO *
C     * THE INSTRUCTION LABELED 20;  OTHERWISE, CONTINUE TO THE NEXT    *
C     * INSTRUCTION.                                                    *
C     *****************************************************************
C
      IF (TEMPX .EQ. -98765.0) GO TO 20
C
C     *****************************************************************
C     * AT THIS POINT, N INDICATES THE NUMBER OF THE DATA POINT JUST    *
C     * READ AND STORED IN TEMPX.  IT WILL BE FOUND CONVENIENT TO STORE *
C     * THE NTH DATA POINT IN THE NTH ELEMENT OF THE VECTOR CALLED X.   *
C     * NOTE THAT X(N) INDICATES THE NTH ELEMENT OF THE VECTOR CALLED X.*
C     *****************************************************************
C
      X(N) = TEMPX
C
C     *****************************************************************
C     * ADD ONE TO THE COUNTER AND SUBSCRIPT (N) TO INDICATE THE NUMBER *
C     * OF THE NEXT DATA POINT ABOUT TO BE PROCESSED AND THE NEXT ELE-  *
C     * MENT OF THE VECTOR ABOUT TO BE USED.                           *
C     *****************************************************************
C
      N = N + 1
C
C     *****************************************************************
C     * TEST THE COUNTER AND SUBSCRIPT (N) TO DETERMINE IF THE NUMBER OF*
C     * THE NEXT DATA POINT IS GREATER THAN 300;  IF IT IS, THE PROGRAM *
C     * STOPS.  NOTE THAT THE CHOICE OF 300 AS A LIMIT IS AN ARBITRARY  *
C     * RESTRICTION BUT IT MUST BE ENFORCED SINCE THE VECTOR X CAN HOLD *
C     * A MAXIMUM OF 300 DATA POINTS.                                   *
C     *****************************************************************
C
      IF (N.GT.300) STOP
C
C     *****************************************************************
C     * BRANCH BACK TO GET THE NEXT DATA POINT.                         *
C     *****************************************************************
C
      GO TO 10
C
C     *****************************************************************
C     * SINCE THE SPECIAL VALUE OF -98765 WAS COUNTED ABOVE, AND SINCE  *
C     * IT IS NOT TO BE CONSIDERED AS ONE OF THE DATA POINTS IN THE SER-*
C     * IES, THE NUMBER (N) MUST NOW BE REDUCED BY ONE.                 *
C     *****************************************************************
C
 20   N = N - 1
```

```
C
C
C     **********************************************************************
C     * IF THE NUMBER (N) IS EQUAL TO ZERO THE MEDIAN CANNOT BE CALCULA-*
C     * TED, SO THE PROGRAM STOPS.                                      *
C     **********************************************************************
C
      IF (N.EQ.0) STOP
C
C     **********************************************************************
C     * STEP 3.  FIRST DETERMINE IF THE NUMBER (N) OF DATA POINTS IN THE*
C     * SERIES IS EVEN OR ODD.                                          *
C     **********************************************************************
C
C
C     **********************************************************************
C     * DIVIDING THE INTERGER VARIABLE N BY THE INTEGER CONSTANT 2 WILL *
C     * TRUNCATE THE FRACTIONAL PORTION OF THE RESULT (J) IF ONE IS     *
C     * PRESENT (THE RESULT (J) WOULD CONTAIN A FRACTIONAL PART IF N     *
C     * WERE ODD).                                                      *
C     **********************************************************************
C
      J = N / 2
C
C     **********************************************************************
C     * MULTIPLY THE RESULT (J) FROM THE ABOVE BY 2.                    *
C     **********************************************************************
C
      J = J * 2
C
C     **********************************************************************
C     * IF THE NUMBER (N) IS EVEN THEN J AND N SHOULD BE EQUAL;  IF THEY*
C     * ARE EQUAL, BRANCH TO THE ROUTINE WHICH CALCULATES THE MEDIAN FOR*
C     * AN EVEN NUMBER (N) OF DATA POINTS. OTHERWISE CONTINUE TO THE    *
C     * NEXT INSTRUCTION, ASSUMING THAT THE NUMBER (N) IS ODD.          *
C     **********************************************************************
C
      IF (J.EQ.N) GO TO 25
C
C     **********************************************************************
C     * CALCULATE THE SUBSCRIPT (INDEX) WHICH WILL INDICATE THE PARTICU-*
C     * LAR ELEMENT OF THE VECTOR (X) THAT IS THE MEDIAN.  THIS IS DONE *
C     * BY DIVIDING THE NUMBER (N) OF DATA POINTS PLUS 1 BY 2.  SINCE N *
C     * AT THIS POINT IS ODD, THE RESULT OF THE DIVISION WILL BE A WHOLE*
C     * NUMBER.                                                         *
C     **********************************************************************
C
      INDEX = (N + 1) / 2
C
C     **********************************************************************
C     * DETERMINE THE MEDIAN (XMED) OF THE SERIES BY SETTING IT EQUAL TO*
C     * THE PARTICULAR ELEMENT OF THE VECTOR INDICATED BY INDEX.        *
C     **********************************************************************
C
      XMED = X(INDEX)
C
C     **********************************************************************
C     * BRANCH TO THE STATEMENT LABELED 30 TO PRINT OUT THE MEDIAN.     *
C     **********************************************************************
C
      GO TO 30
C
C     **********************************************************************
C     * STEP 4.  AT THIS POINT IT HAS BEEN DETERMINED THAT THE NUMBER   *
C     * (N) IS EVEN.                                                    *
```

```
C      ********************************************************************
C
C
C      ********************************************************************
C      *  CALCULATE THE FIRST SUBSCRIPT (INDEX1) WHICH WILL INDICATE THE  *
C      *  FIRST OF THE ELEMENTS OF THE VECTOR (X) TO BE USED IN DETERMIN- *
C      *  ING THE MEDIAN.  THIS IS DONE BY DIVIDING THE NUMBER (N) BY 2,  *
C      *  AND SINCE N IS EVEN, THE RESULT WILL BE A WHOLE NUMBER.         *
C      ********************************************************************
C
  25   INDEX1 = N / 2
C
C      ********************************************************************
C      *  CALCULATE THE SECOND SUBSCRIPT (INDEX2) AS ABOVE, BUT ADD 1 TO  *
C      *  THE RESULT.                                                     *
C      ********************************************************************
C
       INDEX2 = (N / 2) + 1
C
C      ********************************************************************
C      *  DETERMINE THE SUM (SUM) OF THE TWO ELEMENTS OF THE VECTOR (X)   *
C      *  INDICATED BY INDEX1 AND INDEX2.                                 *
C      ********************************************************************
C
       SUM = X(INDEX1) + X(INDEX2)
C
C      ********************************************************************
C      *  CALCULATE THE MEDIAN (XMED) BY DIVIDING THE RESULT (SUM) ABOVE  *
C      *  BY 2.                                                           *
C      ********************************************************************
C
       XMED = SUM / 2.0
C
C      ********************************************************************
C      *  PRINT OUT THE MEDIAN (XMED) OF THE SERIES.                      *
C      ********************************************************************
C
  30   WRITE (6, 600) XMED
 600   FORMAT (1X,28HTHE MEDIAN OF THE SERIES IS , F10.3)
C
C      ********************************************************************
C      *  STOP THE PROGRAM.                                               *
C      ********************************************************************
C
       STOP
C
C      ********************************************************************
C      *  INDICATE THE END OF THE SOURCE PROGRAM.                         *
C      ********************************************************************
C
       END
```

D. Example Problems

Example 1   Using the preceding program, calculate the median for the following sequences of numbers:

(a) 2, 4, 8, 12, 22, 28

(b) 1, 3, 5, 7, 9

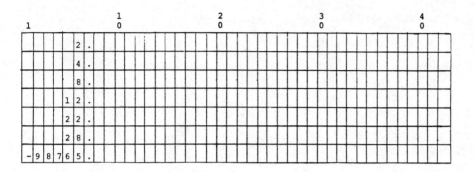

Input Data

THE MEDIAN OF THE SERIES IS 10.000

Solution (a)

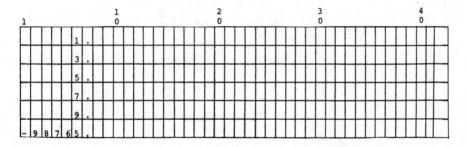

THE MEDIAN OF THE SERIES IS 5.000

Solution (b)

A special class achieved the following test scores: 70, 77, 83, 87, 91, 93, 99. What is the median score?

Example 2

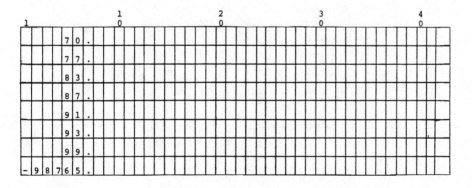

Input Data

THE MEDIAN OF THE SERIES IS 87.000

Solution

☆ Calculation of the Mean Deviation for Ungrouped Data

A. Statement of Problem

The mean deviation (M.D.) for a set of N numbers, $X_1, X_2, \ldots, X_N$, with mean m, is defined as:

$$\text{M.D.} = \frac{\displaystyle\sum_{i=1}^{N} |X_i - m|}{N}$$

Write a program to calculate the mean deviation for a set of N data points.

B. Algorithm

1. Determine the number (N) of data points·(X) in the series for which the mean deviation is desired.
2. Insure that the number (N) of data points (X) to be used is 1 or more.
3. Calculate the arithmetic mean (AVG) of the series of data points.
   a. Accumulate the sum of the data points (SUM) for the entire series of N numbers.
   b. Calculate the arithmetic mean (AVG) of the series by dividing the sum of the data points (SUM) by the number (N) of data points in the series.
4. Accumulate the sum of the absolute deviations from the mean (SUMDEV) of all N numbers in the series.
   a. Determine the deviation from the mean (DEVFRM) for a data point (X) by subtracting the arithmetic mean (AVG) of the series from each data point (X).
   b. Determine the absolute value (ABSDEV) of the deviation from the mean (DEVFRM) obtained in Step 4a.
   c. Accumulate the sum of the absolute deviations from the mean (SUMDEV) by summing the absolute values (ABSDEV) obtained in Step 4b.
5. Calculate the mean deviation (DEVM) by dividing the sum of the absolute deviations from the mean (SUMDEV) by the number (N) of data points in the series.

C. General Program

```
C
C
C      *****************************************************************
C      * THIS PROGRAM COMPUTES THE MEAN DEVIATION OF A SERIES.         *
C      *****************************************************************
C
C
C      *****************************************************************
C      * INFORM THE COMPUTER THAT X WILL BE A VECTOR OF DATA POINTS WITH *
C      * A MAXIMUM OF 300 ELEMENTS.  THIS VECTOR WILL BE USED TO STORE  *
C      * THE DATA POINTS IN THE SERIES FOR FUTURE REFERENCE.  NOTE THAT *
C      * THE CHOICE OF 300 AS A LIMIT IS ARBITRARY.                     *
C      *****************************************************************
C
       DIMENSION X(300)
C
C      *****************************************************************
C      * STEP 1.  DETERMINE THE NUMBER (N) OF DATA POINTS (X) IN THE SER-*
C      * IES.                                                           *
C      *****************************************************************
C
       READ (5, 500) N
  500  FORMAT (I6)
C
C      *****************************************************************
C      * STEP 2.  TEST THE NUMBER (N) OF DATA POINTS (X) IN THE SERIES TO*
C      * INSURE IT IS 1 OR MORE.  IF IT IS NOT, THE PROGRAM STOPS.      *
C      *****************************************************************
C
       IF (N .LT. 1) STOP
C
C      *****************************************************************
C      * TEST THE NUMBER (N) OF DATA POINTS (X) IN THE SERIES TO INSURE *
C      * IT IS 300 OR LESS.  IF IT IS NOT, THE PROGRAM STOPS.  NOTE THAT*
C      * THE CHOICE OF 300 IS AN ARBITRARY RESTRICTION OF THE PROGRAM;  *
C      * THIS RESTRICTION MUST BE ENFORCED SINCE THE VECTOR X CAN HOLD A *
C      * MAXIMUM OF 300 DATA POINTS.                                    *
C      *****************************************************************
C
       IF (N .GT. 300) STOP
C
C      *****************************************************************
C      * STEP 3.  CALCULATE THE ARITHMETIC MEAN (AVG) OF THE SERIES.    *
C      *****************************************************************
C
C
C      *****************************************************************
C      * INITIALIZE THE ACCUMULATOR OF THE SUM OF THE DATA POINTS (SUM) *
C      * BY SETTING IT EQUAL TO ZERO.                                   *
C      *****************************************************************
C
       SUM = 0.0
C
C      *****************************************************************
C      * INITIALIZE THE COUNTER AND SUBSCRIPT (K) BY SETTING IT EQUAL TO *
C      * ONE.  THIS WILL BE USED TO INDICATE THE NUMBER OF THE DATA POINT*
C      * (X) ABOUT TO BE PROCESSED; IT WILL ALSO BE USED AS A SUBSCRIPT  *
C      * TO REFERENCE ANY PARTICULAR DATA POINT IN THE VECTOR.          *
C      *****************************************************************
C
       K = 1
```

```
C
C      ****************************************************************
C      * GET THE KTH DATA POINT.  AT THIS POINT, K INDICATES THE NUMBER *
C      * OF THE DATA POINT (X) ABOUT TO BE READ.  IT WILL BE FOUND CON- *
C      * VENIENT TO READ THE KTH DATA POINT INTO THE KTH ELEMENT OF THE  *
C      * VECTOR.  NOTE THAT THE KTH ELEMENT OF THE VECTOR IS INDICATED BY*
C      * X(K).                                                          *
C      ****************************************************************
C
    7    READ (5, 501) X(K)
  501    FORMAT (F10.3)
C
C      ****************************************************************
C      * STEP 3(A).  ACCUMULATE THE SUM OF THE DATA POINTS (SUM) BY ADD- *
C      * ING THE VALUE OF THE KTH DATA POINT (X(K)) TO THE PREVIOUS VALUE*
C      * OF THE SUM OF THE DATA POINTS (SUM).                           *
C      ****************************************************************
C
         SUM = SUM + X(K)
C
C      ****************************************************************
C      * ADD ONE TO THE COUNTER AND SUBSCRIPT (K) TO INDICATE THE NUMBER *
C      * OF THE NEXT DATA POINT ABOUT TO BE PROCESSED AND THE NEXT ELE- *
C      * MENT OF THE VECTOR ABOUT TO BE USED.                          *
C      ****************************************************************
C
         K = K + 1
C
C      ****************************************************************
C      * IF ALL THE DATA POINTS HAVE NOT BEEN PROCESSED, BRANCH BACK TO *
C      * GET ANOTHER; OTHERWISE CONTINUE TO THE NEXT INSTRUCTION.       *
C      ****************************************************************
C
         IF (K .LE. N) GO TO 7
C
C      ****************************************************************
C      * STEP 3(B).  CALCULATE THE ARITHMETIC MEAN (AVG) OF THE SERIES BY*
C      * DIVIDING THE SUM OF THE DATA POINTS (SUM) BY THE NUMBER (N) OF  *
C      * DATA POINTS IN THE SERIES.                                     *
C      ****************************************************************
C
         AVG = SUM / N
C
C      ****************************************************************
C      * STEP 4.  ACCUMULATE THE SUM OF THE ABSOLUTE DEVIATIONS FROM THE *
C      * MEAN (SUMDEV) FOR ALL N DATA POINTS IN THE SERIES.  NOTE THAT   *
C      * THE SERIES OF DATA POINTS HAS CONVENIENTLY BEEN STORED IN THE   *
C      * VECTOR SO THAT THE DATA POINTS MAY BE REFERENCED AGAIN IN THIS  *
C      * STEP.                                                          *
C      ****************************************************************
C
C
C
C      ****************************************************************
C      * INITIALIZE THE ACCUMULATOR OF THE SUM OF THE ABSOLUTE DEVIATIONS*
C      * FROM THE MEAN (SUMDEV) BY SETTING IT EQUAL TO ZERO.            *
C      ****************************************************************
C
         SUMDEV = 0.0
C
C      ****************************************************************
C      * REINITIALIZE THE COUNTER AND SUBSCRIPT (K) BY SETTING IT EQUAL  *
C      * TO ONE.  ITS USE HERE WILL BE THE SAME AS ABOVE.              *
C      ****************************************************************
C
         K = 1
```

```
C
C
C     *************************************************************
C     * STEP 4(A).  DETERMINE THE DEVIATION FROM THE MEAN (DEVFRM) FOR   *
C     * THE KTH DATA POINT (X(K)) BY SUBTRACTING THE ARITHMETIC MEAN     *
C     * (AVG) FROM THE KTH ELEMENT (X(K)) OF THE VECTOR OF DATA POINTS.  *
C     *************************************************************
C
   20 DEVFRM = X(K) - AVG
C
C     *************************************************************
C     * STEP 4(B).  DETERMINE THE ABSOLUTE VALUE (ABSDEV) OF THE DEVIA-  *
C     * TION FROM THE MEAN (DEVFRM) BY USE OF THE ABSOLUTE VALUE FUNC-   *
C     * TION WHICH IS CALLED ABS.                                       *
C     *************************************************************
C
      ABSDEV = ABS(DEVFRM)
C
C     *************************************************************
C     * STEP 4(C).  ACCUMULATE THE SUM OF THE ABSOLUTE DEVIATIONS FROM   *
C     * THE MEAN (SUMDEV) BY ADDING THE VALUE OF THE CURRENT ABSOLUTE    *
C     * DEVIATION FROM THE MEAN (ABSDEV) TO THE PREVIOUS VALUE OF THE    *
C     * SUM OF THE ABSOLUTE DEVIATIONS FROM THE MEAN (SUMDEV).           *
C     *************************************************************
C
      SUMDEV = SUMDEV + ABSDEV
C
C     *************************************************************
C     * ADD ONE TO THE COUNTER AND SUBSCRIPT (K) TO INDICATE THE NEXT    *
C     * ELEMENT OF THE VECTOR OF DATA POINTS TO BE PROCESSED.            *
C     *************************************************************
C
      K = K + 1
C
C     *************************************************************
C     * IF ALL THE ELEMENTS OF THE VECTOR OF DATA POINTS HAVE NOT BEEN   *
C     * PROCESSED, BRANCH BACK TO BEGIN PROCESSING THE NEXT.  OTHERWISE  *
C     * CONTINUE TO THE NEXT INSTRUCTION.                               *
C     *************************************************************
C
      IF (K .LE. N) GO TO 20
C
C     *************************************************************
C     * STEP 5.  CALCULATE THE MEAN DEVIATION (DEVM) OF THE SERIES BY    *
C     * DIVIDING THE SUM OF THE ABSOLUTE DEVIATIONS FROM THE MEAN        *
C     * (SUMDEV) BY THE NUMBER (N) OF DATA POINTS IN THE SERIES.         *
C     *************************************************************
C
      DEVM = SUMDEV / N
C
C     *************************************************************
C     * PRINT OUT THE MEAN DEVIATION OF THE SERIES (DEVM).              *
C     *************************************************************
C
      WRITE (6, 600) DEVM
  600 FORMAT (1X,36HTHE MEAN DEVIATION OF THE SERIES IS , F10.3)
C
C     *************************************************************
C     * STOP THE PROGRAM.                                              *
C     *************************************************************
C
      STOP
C
C     *************************************************************
C     * INDICATE THE END OF THE SOURCE PROGRAM.                        *
C     *************************************************************
C
      END
```

D. Example Problems

Example 1   Ten students indicated their total number of credit hours taken during a given quarter as: 12, 16, 18, 9, 15, 17, 10, 11, 14, and 13. Find the mean deviation of the number of credit hours taken by the ten students.

Input Data

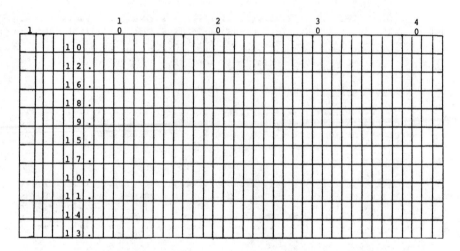

Solution   THE MEAN DEVIATION OF THE SERIES IS 2.500

Example 2   A major cereal producer took a random sample of the weights of one of their sixteen-ounce packages of breakfast cereal. The sample weights were 15.0, 14.8, 16.2, 14.3, 17.0, 15.8, 16.1, 16.0, 15.7, and 16.4. Calculate the mean deviation to determine the extent of the variability of the packages about their mean.

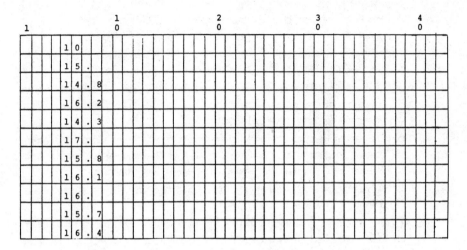

Input Data

THE MEAN DEVIATION OF THE SERIES IS 0.624   *Solution*

☆ Calculation of the Variance of a Set of *N* Numbers

A. Statement of Problem

Given a set of N variates, $X_1, X_2, \ldots, X_N$, whose mean is m. The variance σ^2 is defined as:

$$\sigma^2 = \frac{\displaystyle\sum_{i-1}^{N} (X_i - m)^2}{N}$$

Write a computer program to calculate the variance of a set of N numbers.

B. Algorithm

1. Determine the number (N) of data points (X) in the series for which the variance is desired.
2. Insure that the number (N) of data points (X) to be used is 1 or more.
3. Calculate the arithmetic mean (AVG) of the series of data points.
    a. Accumulate the sum of the data points (SUM) for the entire series of *N* numbers.
    b. Calculate the arithmetic mean (AVG) of the series by divid-

ing the sum of the data points (SUM) by the number (N) of data points in the series.

4. Accumulate the sum of the squared deviations from the mean (SUMDSQ) for all N numbers in the series.

   a. Determine the deviation from the mean (DEVFRM) for a data point (X) by subtracting the arithmetic mean (AVG) of the series from the data point (X).

   b. Determine the squared deviation from the mean (SQDEV) for a data point (X) by multiplying the deviation from the mean (DEVFRM) obtained in step 4a by itself.

   c. Accumulate the sum of the squared deviations from the mean (SUMDSQ).

5. Calculate the variance (VAR) of the series by dividing the sum of the squared deviations from the mean (SUMDSQ) by the number (N) of data points in the series.

C. General Program

```
C
C
C      ****************************************************************
C      *  THIS PROGRAM COMPUTES THE VARIANCE OF A SERIES.            *
C
C      ****************************************************************
C
C
C
C      ****************************************************************
C      *  INFORM THE COMPUTER THAT X WILL BE A VECTOR OF DATA POINTS WITH  *
C      *  A MAXIMUM OF 300 ELEMENTS.  THIS VECTOR WILL BE USED TO STORE    *
C      *  THE DATA POINTS IN THE SERIES FOR FUTURE REFERENCE.  NOTE THAT   *
C      *  THE CHOICE OF 300 AS A LIMIT IS ARBITRARY.                       *
C      ****************************************************************
C
       DIMENSION X(300)
C
C      ****************************************************************
C      *  STEP 1.  DETERMINE THE NUMBER (N) OF DATA POINTS (X) IN THE SER-*
C      *  IES.                                                            *
C      ****************************************************************
C
       READ (5, 500) N
  500  FORMAT (I6)
C
C      ****************************************************************
C      *  STEP 2.  TEST THE NUMBER (N) OF DATA POINTS (X) IN THE SERIES TO*
C      *  INSURE IT IS 1 OR MORE.  IF IT IS NOT, THE PROGRAM STOPS.       *
C      ****************************************************************
C
       IF (N .LT. 1) STOP
C
C      ****************************************************************
C      *  TEST THE NUMBER (N) OF DATA POINTS (X) IN THE SERIES TO INSURE   *
C      *  IT IS 300 OR LESS.  IF IT IS NOT, THE PROGRAM STOPS.  NOTE THAT  *
C      *  THE CHOICE OF 300 IS AN ARBITRARY RESTRICTION OF THE PROGRAM;    *
C      *  THIS RESTRICTION MUST BE ENFORCED SINCE THE VECTOR X CAN HOLD A  *
C      *  MAXIMUM OF 300 DATA POINTS.                                      *
```

```
C     ************************************************************
C
      IF (N .GT. 300) STOP
C
C     ************************************************************
C     * STEP 3.  CALCULATE THE ARITHMETIC MEAN (AVG) OF THE SERIES.    *
C     ************************************************************
C
C
C     ************************************************************
C     * INITIALIZE THE ACCUMULATOR OF THE SUM OF THE DATA POINTS (SUM) *
C     * BY SETTING IT EQUAL TO ZERO.                                   *
C     ************************************************************
C
      SUM = 0.0
C
C     ************************************************************
C     * INITIALIZE THE COUNTER AND SUBSCRIPT (K) BY SETTING IT EQUAL TO *
C     * ONE.  THIS WILL BE USED TO INDICATE THE NUMBER OF THE DATA POINT*
C     * (X) ABOUT TO BE PROCESSED; IT WILL ALSO BE USED AS A SUBSCRIPT  *
C     * TO REFERENCE ANY PARTICULAR DATA POINT IN THE VECTOR.           *
C     ************************************************************
C
      K = 1
C
C     ************************************************************
C     * GET THE KTH DATA POINT.  AT THIS POINT, K INDICATES THE NUMBER  *
C     * OF THE DATA POINT (X) ABOUT TO BE READ.  IT WILL BE FOUND CON-  *
C     * VENIENT TO READ THE KTH DATA POINT INTO THE KTH ELEMENT OF THE  *
C     * VECTOR.  NOTE THAT THE KTH ELEMENT OF THE VECTOR IS INDICATED BY*
C     * X(K).                                                           *
C     ************************************************************
C
   12 READ (5, 501) X(K)
  501 FORMAT (F10.3)
C
C     ************************************************************
C     * STEP 3(A).  ACCUMULATE THE SUM OF THE DATA POINTS (SUM) BY ADD- *
C     * ING THE VALUE OF THE KTH DATA POINT (X(K)) TO THE PREVIOUS VALUE*
C     * OF THE SUM OF THE DATA POINTS (SUM).                            *
C     ************************************************************
C
      SUM = SUM + X(K)
C
C     ************************************************************
C     * ADD ONE TO THE COUNTER AND SUBSCRIPT (K) TO INDICATE THE NUMBER *
C     * OF THE NEXT DATA POINT ABOUT TO BE PROCESSED AND THE NEXT ELE-  *
C     * MENT OF THE VECTOR ABOUT TO BE USED.                            *
C     ************************************************************
C
      K = K + 1
C
C     ************************************************************
C     * IF ALL THE DATA POINTS HAVE NOT BEEN PROCESSED, BRANCH BACK TO  *
C     * GET ANOTHER; OTHERWISE CONTINUE TO THE NEXT INSTRUCTION.        *
C     ************************************************************
C
      IF (K .LE. N) GO TO 12
C
C     ************************************************************
C     * STEP 3(B).  CALCULATE THE ARITHMETIC MEAN (AVG) OF THE SERIES BY*
C     * DIVIDING THE SUM OF THE DATA POINTS (SUM) BY THE NUMBER (N) OF   *
C     * DATA POINTS IN THE SERIES.                                      *
```

```
C     ************************************************************************
C
C           AVG = SUM / N
C
C     ************************************************************************
C     * STEP 4.  ACCUMULATE THE SUM OF THE SQUARED DEVIATIONS FROM THE      *
C     * MEAN (SUMDSQ) FOR ALL N NUMBERS IN THE SERIES.                      *
C     ************************************************************************
C
C
C     ************************************************************************
C     * INITIALIZE THE ACCUMULATOR OF THE SUM OF THE SQUARED DEVIATIONS     *
C     * (SUMDSQ) BY SETTING IT EQUAL TO ZERO.                               *
C     ************************************************************************
C
      SUMDSQ = 0.0
C
C     ************************************************************************
C     * REINITIALIZE THE COUNTER AND SUBSCRIPT (K) BY SETTING IT EQUAL      *
C     * TO ONE.  ITS USE HERE IS THE SAME AS ABOVE.                         *
C     ************************************************************************
C
      K = 1
C
C     ************************************************************************
C     * STEP 4(A).  DETERMINE THE DEVIATION FROM THE MEAN (DEVFRM) OF       *
C     * THE KTH DATA POINT BY SUBTRACTING THE ARITHMETIC MEAN OF THE        *
C     * SERIES (AVG) FROM THE KTH ELEMENT OF THE VECTOR.  NOTE THAT X(K)*
C     * DENOTES THE KTH ELEMENT OF THE VECTOR WHICH IS ALSO THE KTH DATA*
C     * POINT.                                                              *
C     ************************************************************************
C
   15 DEVFRM = X(K) - AVG
C
C     ************************************************************************
C     * STEP 4(B).  DETERMINE THE SQUARED DEVIATION FROM THE MEAN           *
C     * (SQDEV) BY      MULTIPLYING THE DEVIATION (DEVFRM) BY ITSELF.       *
C     ************************************************************************
C
      SQDEV = DEVFRM * DEVFRM
C
C     ************************************************************************
C     * STEP 4(C).  ACCUMULATE THE SUM OF THE SQUARED DEVIATIONS FROM       *
C     * THE MEAN (SUMDSQ) BY ADDING THE VALUE OF THE CURRENT SQUARED        *
C     * DEVIATION (SQDEV) TO THE PREVIOUS VALUE OF THE SUM OF THE           *
C     * SQUARED DEVIATIONS FROM THE MEAN (SUMDSQ).                          *
C     ************************************************************************
C
      SUMDSQ = SUMDSQ + SQDEV
C
C     ************************************************************************
C     * ADD ONE TO THE COUNTER AND SUBSCRIPT (K) TO INDICATE THE NEXT       *
C     * ELEMENT OF THE VECTOR OF DATA POINTS TO BE PROCESSED.               *
C     ************************************************************************
C
      K = K + 1
C
C     ************************************************************************
C     * IF ALL THE ELEMENTS OF THE VECTOR OF DATA POINTS HAVE NOT BEEN      *
C     * PROCESSED, BRANCH BACK TO BEGIN PROCESSING THE NEXT.  OTHERWISE     *
C     * CONTINUE TO THE NEXT INSTRUCTION.                                   *
C     ************************************************************************
C
      IF (K .LE. N) GO TO 15
```

```
C
C
C     ***********************************************************************
C     * STEP 5.  CALCULATE THE VARIANCE (VAR) OF THE SERIES BY DIVIDING *
C     * THE SUM OF THE SQUARED DEVIATIONS (SUMDSQ) BY THE NUMBER (N) OF *
C     * DATA POINTS IN THE SERIES.                                      *
C     ***********************************************************************
C
      VAR = SUMDSQ / N
C
C
C     ***********************************************************************
C     * PRINT OUT THE VARIANCE (VAR) OF THE SERIES.                     *
C     ***********************************************************************
C
      WRITE (6, 600) VAR
  600 FORMAT (1X,30HTHE VARIANCE OF THE SERIES IS , F10.3)
C
C
C     ***********************************************************************
C     * STOP THE PROGRAM.                                               *
C     ***********************************************************************
C
      STOP
C
C
C     ***********************************************************************
C     * INDICATE THE END OF THE SOURCE PROGRAM.                         *
C     ***********************************************************************
C
      END
```

D. Example Problems

A test was given to 30 psychology students. The scores were as follows: 73, 67, 94, 87, 42, 91, 85, 79, 83, 87, 84, 79, 78, 77, 75, 83, 85, 37, 98, 86, 78, 88, 72, 90, 80, 70, 69, 78, 83, 85. Find the variance of these scores.

Example 1

Input Data

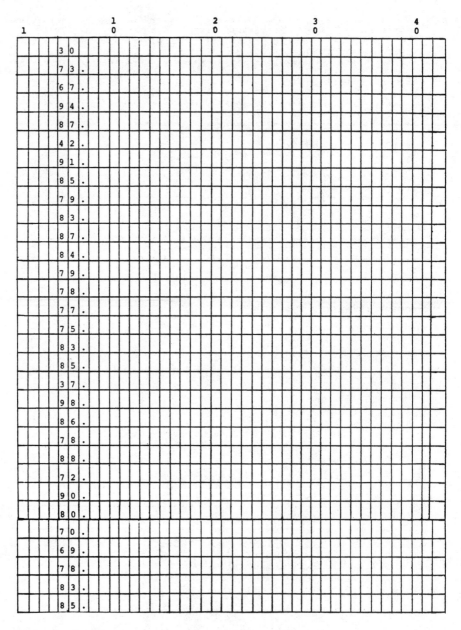

Solution   THE VARIANCE OF THE SERIES IS 161.845

A sample speed check by the state police on an interstate highway   *Example 2*
yielded the following speeds: 58.5, 55.0, 48.0, 62.5, 66.5, 72.5, 63.5,
59.5, 59.0, 62.0, 61.0, 60.5, 60.2, 59.7, 61.0, 58.0. Calculate the
variance of the above speeds.

Input Data

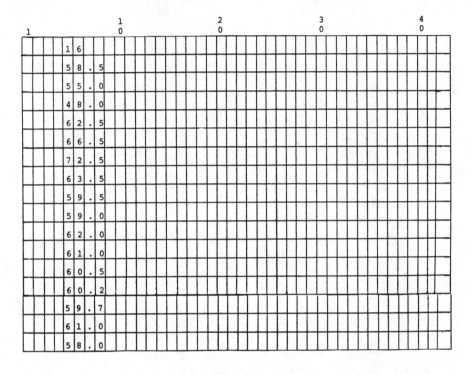

THE VARIANCE OF THE SERIES IS 24.779   *Solution*

☆ Calculation of the Standard Deviation of *N* Numbers

A. Statement of Problem

Given a set of N variates, $X_1, X_2, \ldots, X_N$, whose mean is m. The
standard deviation σ is defined as:

$$\sigma = \sqrt{\dfrac{\displaystyle\sum_{i=1}^{N} (X_i - m)^2}{N}}$$

Write a computer program to calculate the standard deviation of a
set of N numbers.

B. Algorithm

1. Determine the number (N) of data points (X) in the series for which the standard deviation is desired.

2. Insure that the number (N) of data points (X) to be used is 1 or more.

3. Calculate the arithmetic mean (AVG) of the series of data points.

    a. Accumulate the sum of the data points (SUM) for the entire series of N numbers.

    b. Calculate the arithmetic mean (AVG) of the series by dividing the sum of the data points (SUM) by the number (N) of data points in the series.

4. Accumulate the sum of the squared deviations from the mean (SUMDSQ) for all N numbers in the series.

    a. Determine the deviations from the mean (DEVFRM) for a data point by subtracting the arithmetic mean (AVG) of the series from the data point (X).

    b. Determine the squared deviations from the mean (SQDEV) for a data point (X) by multiplying the deviation from the mean (DEVFRM) obtained in Step 4a by itself.

    c. Accumulate the sum of the squared deviations from the mean (SUMDSQ).

5. Calculate the standard deviation (STDDEV) of the series by extracting the square root of the quantity [sum of the squared deviations from the mean (SUMDSQ) divided by the number (N) of data points in the series].

C. General Program

```
C
C
C    *****************************************************************
C    * THIS PROGRAM COMPUTES THE STANDARD DEVIATION OF A SERIES.     *
C    *****************************************************************
C
C
C    *****************************************************************
C    * INFORM THE COMPUTER THAT X WILL BE A VECTOR OF DATA POINTS WITH *
C    * A MAXIMUM OF 300 ELEMENTS.  THIS VECTOR WILL BE USED TO STORE   *
C    * THE DATA POINTS IN THE SERIES FOR FUTURE REFERENCE.  NOTE THAT  *
C    * THE CHOICE OF 300 AS A LIMIT IS ARBITRARY.                      *
C    *****************************************************************
C
     DIMENSION X(300)
C
C
C    *****************************************************************
C    * STEP 1.  DETERMINE THE NUMBER (N) OF DATA POINTS (X) IN THE SER-*
C    * IES.                                                            *
```

```
C     ************************************************************************
C
      READ (5, 500) N
  500 FORMAT (I6)
C
C     ************************************************************************
C     * STEP 2.  TEST THE NUMBER (N) OF DATA POINTS (X) IN THE SERIES TO*
C     * INSURE IT IS 1 OR MORE.  IF IT IS NOT, THE PROGRAM STOPS.       *
C     ************************************************************************
C
      IF (N .LT. 1) STOP
C
C     ************************************************************************
C     * TEST THE NUMBER (N) OF DATA POINTS (X) IN THE SERIES TO INSURE  *
C     * IT IS 300 OR LESS.  IF IT IS NOT, THE PROGRAM STOPS.  NOTE THAT *
C     * THE CHOICE OF 300 IS AN ARBITRARY RESTRICTION OF THE PROGRAM;   *
C     * THIS RESTRICTION MUST BE ENFORCED SINCE THE VECTOR X CAN HOLD A *
C     * MAXIMUM OF 300 DATA POINTS.                                     *
C     ************************************************************************
C
      IF (N .GT. 300) STOP
C
C     ************************************************************************
C     * STEP 3.  CALCULATE THE ARITHMETIC MEAN (AVG) OF THE SERIES.     *
C     ************************************************************************
C
C
C
C     ************************************************************************
C     * INITIALIZE THE ACCUMULATOR OF THE SUM OF THE DATA POINTS (SUM)  *
C     * BY SETTING IT EQUAL TO ZERO.                                    *
C     ************************************************************************
C
      SUM = 0.0
C
C     ************************************************************************
C     * INITIALIZE THE COUNTER AND SUBSCRIPT (K) BY SETTING IT EQUAL TO *
C     * ONE.  THIS WILL BE USED TO INDICATE THE NUMBER OF THE DATA POINT*
C     * (X) ABOUT TO BE PROCESSED; IT WILL ALSO BE USED AS A SUBSCRIPT  *
C     * TO REFERENCE ANY PARTICULAR DATA POINT IN THE VECTOR.           *
C     ************************************************************************
C
      K = 1
C
C     ************************************************************************
C     * GET THE KTH DATA POINT.  AT THIS POINT, K INDICATES THE NUMBER  *
C     * OF THE DATA POINT (X) ABOUT TO BE READ.  IT WILL BE FOUND CON-  *
C     * VENIENT TO READ THE KTH DATA POINT INTO THE KTH ELEMENT OF THE  *
C     * VECTOR.  NOTE THAT THE KTH ELEMENT OF THE VECTOR IS INDICATED BY*
C     * X(K).                                                           *
C     ************************************************************************
C
   12 READ (5, 501) X(K)
  501 FORMAT (F10.3)
C
C     ************************************************************************
C     * STEP 3(A).  ACCUMULATE THE SUM OF THE DATA POINTS (SUM) BY ADD- *
C     * ING THE VALUE OF THE KTH DATA POINT (X(K)) TO THE PREVIOUS VALUE*
C     * OF THE SUM OF THE DATA POINTS (SUM).                            *
C     ************************************************************************
C
      SUM = SUM + X(K)
C
C     ************************************************************************
C     * ADD ONE TO THE COUNTER AND SUBSCRIPT (K) TO INDICATE THE NUMBER *
```

```
C     * OF THE NEXT DATA POINT ABOUT TO BE PROCESSED AND THE NEXT ELE-   *
C     * MENT OF THE VECTOR ABOUT TO BE USED.                             *
C     ****************************************************************
C
      K = K + 1
C
C     ****************************************************************
C     * IF ALL THE DATA POINTS HAVE NOT BEEN PROCESSED, BRANCH BACK TO   *
C     * GET ANOTHER; OTHERWISE CONTINUE TO THE NEXT INSTRUCTION.         *
C     ****************************************************************
C
      IF (K .LE. N) GO TO 12
C
C     ****************************************************************
C     * STEP 3(B).  CALCULATE THE ARITHMETIC MEAN (AVG) OF THE SERIES BY*
C     * DIVIDING THE SUM OF THE DATA POINTS (SUM) BY THE NUMBER (N) OF   *
C     * DATA POINTS IN THE SERIES.                                       *
C     ****************************************************************
C
      AVG = SUM / N
C
C     ****************************************************************
C     * STEP 4.  ACCUMULATE THE SUM OF THE SQUARED DEVIATIONS FROM THE   *
C     * MEAN (SUMDSQ) FOR ALL N NUMBERS IN THE SERIES.                   *
C     ****************************************************************
C
C
C     ****************************************************************
C     * INITIALIZE THE ACCUMULATOR OF THE SUM OF THE SQUARED DEVIATIONS  *
C     * (SUMDSQ) BY SETTING IT EQUAL TO ZERO.                            *
C     ****************************************************************
C
      SUMDSQ = 0.0
C
C     ****************************************************************
C     * REINITIALIZE THE COUNTER AND SUBSCRIPT (K) BY SETTING IT EQUAL   *
C     * TO ONE.  ITS USE HERE IS THE SAME AS ABOVE.                      *
C     ****************************************************************
C
      K = 1
C
C     ****************************************************************
C     * STEP 4(A).  DETERMINE THE DEVIATION FROM THE MEAN (DEVFRM) OF    *
C     * THE KTH DATA POINT BY SUBTRACTING THE ARITHMETIC MEAN OF THE     *
C     * SERIES (AVG) FROM THE KTH ELEMENT OF THE VECTOR.  NOTE THAT X(K)*
C     * DENOTES THE KTH ELEMENT OF THE VECTOR WHICH IS ALSO THE KTH DATA*
C     * POINT.                                                           *
C     ****************************************************************
   15 DEVFRM = X(K) - AVG
C
C     ****************************************************************
C     * STEP 4(B).  DETERMINE THE SQUARED DEVIATION FROM THE MEAN        *
C     * (SQDEV) BY BY MULTIPLYING THE DEVIATION (DEVFRM) BY ITSELF.      *
C     ****************************************************************
C
      SQDEV = DEVFRM * DEVFRM
C
C     ****************************************************************
C     * STEP 4(C).  ACCUMULATE THE SUM OF THE SQUARED DEVIATIONS FROM    *
C     * THE MEAN (SUMDSQ) BY ADDING THE VALUE OF THE CURRENT SQUARED     *
C     * DEVIATION (SQDEV) TO THE PREVIOUS VALUE OF THE SUM OF THE        *
C     * SQUARED DEVIATIONS FROM THE MEAN (SUMDSQ).                       *
```

```
C     ************************************************************
C
      SUMDSQ = SUMDSQ + SQDEV
C
C     ************************************************************
C     * ADD ONE TO THE COUNTER AND SUBSCRIPT (K) TO INDICATE THE NEXT  *
C     * ELEMENT OF THE VECTOR OF DATA POINTS TO BE PROCESSED.          *
C     ************************************************************
C
      K = K + 1
C
C     ************************************************************
C     * IF ALL THE ELEMENTS OF THE VECTOR OF DATA POINTS HAVE NOT BEEN *
C     * PROCESSED, BRANCH BACK TO BEGIN PROCESSING THE NEXT.  OTHERWISE *
C     * CONTINUE TO THE NEXT INSTRUCTION.                              *
C     ************************************************************
C
      IF (K .LE. N) GO TO 15
C
C     ************************************************************
C     * STEP 5.  CALCULATE THE STANDARD DEVIATION (STDDEV) OF THE SERIES*
C     * BY DIVIDING THE SUM OF THE SQUARED DEVIATIONS (SUMDSQ) BY THE  *
C     * NUMBER (N) OF DATA POINTS IN THE SERIES AND THEN EXTRACTING THE *
C     * SQUARE ROOT OF THE RESULT.  THE SQUARE ROOT FUNCTION, CALLED   *
C     * SQRT, CAN BE USED TO DO THIS.                                  *
C     ************************************************************
C
      STDDEV = SQRT(SUMDSQ / N)
C
C     ************************************************************
C     * PRINT OUT THE STANDARD DEVIATION (STDDEV) OF THE SERIES.       *
C     ************************************************************
C
      WRITE (6, 600) STDDEV
  600 FORMAT (1X,40HTHE STANDARD DEVIATION OF THE SERIES IS , F10.3)
C
C     ************************************************************
C     * STOP THE PROGRAM.                                             *
C     ************************************************************
C
      STOP
C
C     ************************************************************
C     * INDICATE THE END OF THE SOURCE PROGRAM.                       *
C     ************************************************************
C
      END
```

D. Example Problems

Given the following data: 14, 23, 17, 19, 22, 12, 11, 17, 19, 23, 29, 34, 2, 18, 27, 7, 19, and 21, find the standard deviation of the above scores.

Example 1

Input Data

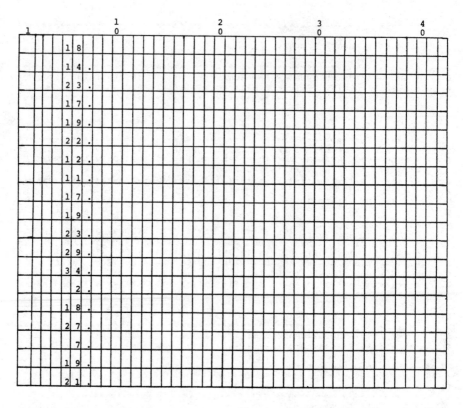

Solution   THE STANDARD DEVIATION OF THE SERIES IS 7.492

Example 2   On the final examination in accounting, 25 students received the following test scores: 78, 83, 70, 75, 92, 83, 77, 75, 64, 62, 65, 95, 50, 89, 90, 67, 61, 81, 69, 62, 54, 85, 79, 81, and 87. Find the standard deviation of the above scores.

Input Data

```
1        1          2          3          4
         0          0          0          0

    2 5
    7 8 .
    8 3 .
    7 0 .
    7 5 .
    9 2 .
    8 3 .
    7 7 .
    7 5 .
    6 4 .
    6 2 .
    6 5 .
    9 5 .
    5 0 .
    8 9 .
    9 0 .
    6 7 .
    6 1 .
    8 1 .
    6 9 .
    6 2 .
    5 4 .
    8 5 .
    7 9 .
    8 1 .
    8 7 .
```

THE STANDARD DEVIATION OF THE SERIES IS 11.881

Solution

☆ Exercises

1. A random sample of 25 students in a statistics class showed the following number of absences: 4, 11, 7, 0, 2, 6, 1, 3, 4, 5, 4, 6, 0, 1, 1, 1, 3, 5, 4, 0, 1, 2, 5. Find the mean number of absences for this class.

2. Fourteen fishermen reported the number of perch caught in Lake Erie. There were 10, 6, 14, 8, 4, 11, 5, 3, 3, 1, 11, 9, 10, 2. Find the median number of fish caught.

3. In the production at Steel City Manufacturing, weights in grams for the production run for ⅝ in. steel nuts were 2.41, 2.49, 2.39, 2.54, 2.45, 2.40, 2.42, 2.47. Find the mean deviation of weights for the above production run.

4. The number of typing errors for five prospective secretaries were observed as: 5, 10, 7, 4, and 6. Find the variance among the observed number of typing errors.

5. The annual salaries for five automobile salesmen were reported as: $8,250, $9,000, $12,500, $11,200, and $10,400. Find the standard deviation among these salaries.

☆ Calculation of the Arithmetic Mean for Grouped Data

A. Statement of the Problem

Suppose we are given a frequency table consisting of M classes with class marks $X_1, X_2, \ldots, X_M$ having respective frequencies $F_1, F_2, \ldots, F_M$. The class mark is determined by the average of the lower bound and upper bound of any one class. Write a computer program to compute the mean of the frequency table, using the formula

$$m = \frac{\displaystyle\sum_{i=1}^{M} X_i F_i}{\displaystyle\sum_{i=1}^{M} F_i}$$

B. Algorithm

1. Determine the number (M) of lower and upper class bounds (XLOWER, XUPPER) and associated frequencies (F) of the frequency table for which the arithmetic mean is desired.

2. Insure that the number (M) of lower and upper class bounds

(XLOWER, XUPPER) and associated frequencies (F) is 1 or more.

3. Calculate the arithmetic mean (AVG) of the frequency table.

   a. Calculate the midpoint of each class (X) by summing the upper bound (XUPPER) and the lower (XLOWER) and dividing the sum by 2.0.

   b. Accumulate the sum of the products (XF) of the class midpoints and frequencies.

   c. Accumulate the sum of the frequencies (SUMF) for all classes in the table.

   d. Calculate the arithmetic mean (AVG) of the frequency table by dividing the sum of the products (XF) of class midpoints and associated frequencies by the sum of the frequencies (SUMF) in the table.

C. General Program

```
C
C     *****************************************************************
C     * THIS PROGRAM COMPUTES THE ARITHMETIC MEAN OF A FREQUENCY TABLE. *
C     *****************************************************************
C
C
C
C     *****************************************************************
C     * STEP 1.  DETERMINE THE NUMBER (M) OF LOWER AND UPPER CLASS      *
C     * BOUNDS (XLOWER, XUPPER) AND ASSOCIATED FREQUENCIES (F) OF THE   *
C     * FREQUENCY TABLE.                                                *
C     *****************************************************************
C
      READ (5, 500) M
  500 FORMAT (I6)
C
C     *****************************************************************
C     * STEP 2.  TEST THE NUMBER (M) OF LOWER AND UPPER CLASS BOUNDS    *
C     * (XLOWER, XUPPER) AND ASSOCIATED FREQUENCIES (F) TO INSURE IT IS *
C     * 1 OR MORE.  IF IT IS NOT, THE PROGRAM STOPS.                    *
C     *****************************************************************
C
      IF (M .LT. 1) STOP
C
C     *****************************************************************
C     * STEP 3.  CALCULATE THE ARITHMETIC MEAN (AVG) OF THE FREQUENCY   *
C     * TABLE.                                                          *
C     *****************************************************************
C
C
C
C     *****************************************************************
C     * INITIALIZE THE COUNTER (J) BY SETTING IT EQUAL TO ZERO.  THIS   *
C     * WILL BE USED TO COUNT THE NUMBER OF CLASS BOUNDS AND FREQUENCIES*
C     * (XLOWER, XUPPER, F) THAT HAVE BEEN PROCESSED.                   *
C     *****************************************************************
C
      J = 0
C
C     *****************************************************************
C     * INITIALIZE THE ACCUMULATOR OF THE PRODUCTS (XF) OF CLASS MID-   *
```

```
C      * POINTS (X) AND ASSOCIATED FREQUENCIES (F) BY SETTING IT EQUAL TO*
C      * ZERO.                                                           *
C      ****************************************************************
C
       XF = 0.0
C
C      ****************************************************************
C      * INITIALIZE THE ACCUMULATOR OF THE SUM OF THE FREQUENCIES (SUMF) *
C      * BY SETTING IT EQUAL TO ZERO.                                    *
C      ****************************************************************
C
       SUMF = 0.0
C
C      ****************************************************************
C      * GET THE NEXT LOWER BOUND, UPPER BOUND, AND FREQUENCY (XLOWER,   *
C      * XUPPER, F).                                                     *
C      ****************************************************************
C
   10  READ (5, 501) XLOWER, XUPPER, F
  501  FORMAT (3F10.3)
C
C      ****************************************************************
C      * STEP 3(A).  CALCULATE THE MIDPOINT (X) OF THIS CLASS BY SUMMING *
C      * ITS BOUNDS (XLOWER, XUPPER) AND DIVIDING THE SUM BY 2.0.        *
C      ****************************************************************
C
       X = (XUPPER + XLOWER) / 2.0
C
C      ****************************************************************
C      * STEP 3(B).  ACCUMULATE THE SUM OF THE PRODUCTS (XF) OF THE CLASS*
C      * MIDPOINT (X) AND ITS ASSOCIATED FREQUENCY (F) BY ADDING TO THE  *
C      * PREVIOUS SUM OF THE PRODUCTS (XF) THE RESULT OF THE MULTIPLICA-  *
C      * TION OF THE CURRENT MIDPOINT (X) BY ITS ASSOCIATED FREQUENCY    *
C      * (F).                                                            *
C      ****************************************************************
C
       XF = XF + (X * F)
C
C      ****************************************************************
C      * STEP 3(C).  ACCUMULATE THE SUM OF THE FREQUENCIES (SUMF) BY ADD-*
C      * ING THE VALUE OF THE CURRENT FREQUENCY (F) TO THE PREVIOUS SUM  *
C      * OF THE FREQUENCIES (SUMF).                                      *
C      ****************************************************************
C
       SUMF = SUMF + F
C
C      ****************************************************************
C      * ADD 1 TO THE COUNTER (J) TO INDICATE THAT ANOTHER SET OF BOUNDS *
C      * AND ASSOCIATED FREQUENCY (XLOWER, XUPPER, F) HAS BEEN PROCESSED.*
C      ****************************************************************
C
       J = J + 1
C
C      ****************************************************************
C      * IF ALL SETS OF BOUNDS AND FREQUENCIES (XLOWER, XUPPER, F) HAVE  *
C      * NOT BEEN PROCESSED, BRANCH BACK TO GET ANOTHER; OTHERWISE, CON-  *
C      * TINUE TO THE NEXT INSTRUCTION.                                  *
C      ****************************************************************
C
       IF (J .LT. M) GO TO 10
C
C      ****************************************************************
C      * STEP 3(D).  CALCULATE THE ARITHMETIC MEAN (AVG) OF THE FREQUENCY*
C      * TABLE BY DIVIDING THE SUM OF THE PRODUCTS (XF) OF MIDPOINTS (X) *
C      * AND FREQUENCIES (F) BY THE SUM OF THE FREQUENCIES (SUMF) OF THE *
C      * TABLE.                                                          *
```

```
C
C      ***********************************************************************
C
       AVG = XF / SUMF
C
C      ***********************************************************************
C      * PRINT OUT THE ARITHMETIC MEAN (AVG) OF THE FREQUECY TABLE.         *
C      ***********************************************************************
C
       WRITE (6, 600) AVG
 600   FORMAT (1X,46HTHE ARITHMETIC MEAN OF THE FREQUENCY TABLE IS ,
      *    F10.3)
C
C      ***********************************************************************
C      * STOP THE PROGRAM.                                                  *
C
C      ***********************************************************************
C
       STOP
C
C      ***********************************************************************
C      * INDICATE THE END OF THE SOURCE PROGRAM.                            *
C      ***********************************************************************
C
       END
```

D. Example Problems

Given the following grouped data:

Example 1

CB	f
15–25	3
25–35	14
35–45	18
45–55	12
55–65	10
65–75	5

find the mean of the above data.

Input Data

Solution   THE ARITHMETIC MEAN OF THE FREQUENCY TABLE is 44.355

Example 2   Given the following grouped data:

CB	f
20–30	2
30–40	5
40–50	8
50–60	3
60–70	2
70–80	5
80–90	2
90–100	1

find the mean of the above data.

Input Data

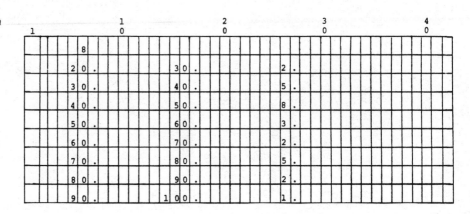

Solution   THE ARITHMETIC MEAN OF THE FREQUENCY TABLE IS 54.286

☆ Calculation of the Mean Deviation for Grouped Data

A. Statement of the Problem

The mean deviation (M.D.) in a frequency table having class midpoints $X_1, X_2, \ldots, X_M$ with respective frequencies $F_1, F_2, \ldots, F_M$ is defined as:

$$\text{M.D.} = \frac{\sum\limits_{i=1}^{M} |X_i - m| F_i}{N}$$

Note that m is the mean of the grouped data and that

$$N = \sum_{i=1}^{M} F_i$$

Write a program to calculate the mean deviation for a set of variates grouped into M classes.

B. Algorithm

1. Determine the number (M) of lower and upper class bounds (XLOWER, XUPPER) and associated frequencies (F) of the frequency table for which the mean deviation is desired.

2. Insure that the number (M) of lower and upper class bounds (XLOWER, XUPPER) and associated frequencies (F) is 1 or more.

3. Calculate the arithmetic mean (AVG) of the frequency table.

   a. Calculate the midpoint of each class (X) by summing the upper and lower bounds (XUPPER, XLOWER) and dividing by 2.0.

   b. Accumulate the sum of the products (XF) of the class mid-points (X) and their associated frequencies (F) for all classes and frequencies in the table.

   c. Accumulate the sum of the frequencies (SUMF) for all classes in the table.

   d. Calculate the arithmetic mean (AVG) of the frequency table by dividing the sum of the products (XF) of midpoints and associated frequencies (SUMF) in the table.

4. Calculate the numerator in the formula (SUMDXF). This is accomplished by summing the product of the absolute deviation of the midpoints (X) from the arithmetic mean (AVG) times the associated frequencies in the table.

   a. Determine the absolute deviation (ABSDEV) of the class midpoint (X) from the arithmetic mean (AVG) by subtracting the arithmetic mean (AVG) from the class midpoint (X) and taking the absolute value of the result.

b. Determine the absolute deviation from the mean times the associated frequency (DEVXF) by multiplying the absolute deviation (ABSDEV) obtained in Step 4a by the associated frequency (F) of the class midpoint.

c. Calculate the numerator (SUMDXF) in the formula by accumulating the sum of the absolute deviations times frequencies (DEVXF) obtained in Step 4b.

5. Calculate the mean deviation (DEVMF) of the table by dividing the numerator of the formula (SUMDXF) by the sum of the frequencies (SUMF) obtained in Step 3c.

C. General Program

```
C
C       ******************************************************************
C       * THIS PROGRAM COMPUTES THE MEAN DEVIATION OF A FREQUENCY TABLE. *
C       ******************************************************************
C
C
C       ******************************************************************
C       * INFORM THE COMPUTER THAT X AND F WILL BE VECTORS OF DATA POINTS *
C       * WITH A MAXIMUM OF 300 ELEMENTS.  THESE VECTORS WILL BE USED TO  *
C       * STORE THE CLASS MIDPOINTS AND ASSOCIATED FREQUENCIES OF THE TA- *
C       * BLE FOR FUTURE REFERENCE.  NOTE THAT THE CHOICE OF 300 AS A LIM-*
C       * IT IS ARBITRARY.                                                *
C       ******************************************************************
C
        DIMENSION X(300), F(300)
C
C       ******************************************************************
C       * STEP 1.  DETERMINE THE NUMBER (M) OF LOWER AND UPPER CLASS      *
C       * BOUNDS (XLOWER, XUPPER) AND ASSOCIATED FREQUENCIES (F) IN THE   *
C       * TABLE.                                                          *
C       ******************************************************************
C
        READ (5, 500) M
  500   FORMAT (I6)
C
C       ******************************************************************
C       * STEP 2.  TEST THE NUMBER (M) OF LOWER AND UPPER CLASS BOUNDS    *
C       * (XLOWER, XUPPER) AND ASSOCIATED FREQUENCIES (F) TO INSURE IT IS *
C       * 1 OR MORE; IF IT IS NOT, THE PROGRAM STOPS.                     *
C       ******************************************************************
C
        IF (M.LT.1) STOP
C
C       ******************************************************************
C       * TEST THE NUMBER (M) OF LOWER AND UPPER CLASS BOUNDS (XLOWER,    *
C       * XUPPER) AND ASSOCIATED FREQUENCIES (F) TO INSURE IT IS 300 OR   *
C       * LESS; IF IT IS NOT, THE PROGRAM STOPS.  NOTE THAT 300 IS AN AR- *
C       * BITRARY RESTRICTION OF THE PROGRAM, BUT IT MUST BE ENFORCED     *
C       * SINCE THE VECTORS X AND F CAN CONTAIN A MAXIMUM OF 300 ELEMENTS.*
C       ******************************************************************
C
        IF (M.GT.300) STOP
C
C       ******************************************************************
C       * STEP 3.  CALCULATE THE ARITHMETIC MEAN (AVG) OF THE FREQUENCY   *
C       * TABLE.                                                          *
```

```
C     ****************************************************************
C
C
C     ****************************************************************
C     * INITIALIZE THE COUNTER AND SUBSCRIPT (J) BY SETTING IT EQUAL TO *
C     * ONE.  THIS WILL BE USED TO INDICATE THE NUMBER OF THE SET OF   *
C     * BOUNDS AND ASSOCIATED FREQUENCIES (XLOWER, XUPPER, F) ABOUT TO *
C     * BE PROCESSED.  IT WILL ALSO BE USED AS A SUBSCRIPT TO REFERENCE *
C     * ANY PARTICULAR ELEMENT OF EITHER OF THE VECTORS IN THE PROGRAM. *
C     ****************************************************************
C
      J = 1
C
C     ****************************************************************
C     * INITIALIZE THE ACCUMULATOR OF THE PRODUCTS (XF) OF CLASS MID- *
C     * POINTS AND ASSOCIATED FREQUENCIES BY SETTING IT EQUAL TO ZERO. *
C     ****************************************************************
C
      XF = 0.0
C
C     ****************************************************************
C     * INITIALIZE THE ACCUMULATOR OF THE SUM OF THE FREQUENCIES (SUMF) *
C     * BY SETTING IT EQUAL TO ZERO.                                    *
C     ****************************************************************
C
      SUMF = 0.0
C
C     ****************************************************************
C     * GET THE NEXT SET OF BOUNDS AND ASSOCIATED FREQUENCY (XLOWER,  *
C     * XUPPER, F).  AT THIS POINT, J INDICATES THE NUMBER OF THE SET OF*
C     * BOUNDS AND ASSOCIATED FREQUENCY ABOUT TO BE READ.  IT WILL BE  *
C     * FOUND CONVENIENT TO READ THE JTH FREQUENCY INTO THE JTH ELEMENT *
C     * OF THE VECTOR F.  NOTE THAT THE JTH ELEMENT OF VECTOR F IS INDI-*
C     * CATED BY F(J).                                                 *
C     ****************************************************************
5     READ (5, 501) XLOWER, XUPPER, F(J)
501   FORMAT (3F10.3)
C
C     ****************************************************************
C     * STEP 3(A).  CALCULATE THE MIDPOINT OF THIS CLASS BY SUMMING ITS *
C     * LOWER AND UPPER BOUNDS AND DIVIDING THE RESULT BY 2.  AGAIN IT  *
C     * WILL BE FOUND CONVENIENT TO STORE THE MIDPOINT OF THE JTH CLASS *
C     * IN THE JTH ELEMENT OF THE VECTOR X.                            *
C     ****************************************************************
C
      X(J) = (XLOWER + XUPPER) / 2.0
C
C     ****************************************************************
C     * STEP 3(B).  ACCUMULATE THE SUM OF THE PRODUCTS (XF) OF THE CLASS*
C     * MIDPOINTS AND ASSOCIATED FREQUENCIES BY MULTIPLYING THE MIDPOINT*
C     * BY THE FREQUENCY AND ADDING THE RESULT OF THIS TO THE PREVIOUS *
C     * SUM OF THE PRODUCTS (XF).  NOTE THAT J INDICATES THE NUMBER OF *
C     * THE ELEMENT OF THE VECTORS WHERE THE CURRENT MIDPOINT AND ITS  *
C     * ASSOCIATED FREQUENCY ARE STORED.  THUS X(J) INDICATES THE CUR- *
C     * RENT MIDPOINT AND F(J) INDICATES THE ASSOCIATED FREQUENCY.     *
C     ****************************************************************
C
      XF = XF + (X(J) * F(J))
C
C     ****************************************************************
C     * STEP 3(C).  ACCUMULATE THE SUM OF THE FREQUENCIES (SUMF) BY ADD-*
C     * ING THE VALUE OF THE CURRENT FREQUENCY, INDICATED BY F(J), TO  *
C     * THE PREVIOUS VALUE OF THE SUM OF THE FREQUENCIES (SUMF).       *
```

```
C     *********************************************************************
C
      SUMF = SUMF + F(J)
C
C     *********************************************************************
C     * ADD ONE TO THE COUNTER AND SUBSCRIPT (J) TO INDICATE THE NUMBER  *
C     * OF THE NEXT SET OF BOUNDS AND ASSOCIATED FREQUENCY ABOUT TO BE   *
C     * PROCESSED AND ALSO TO INDICATE THE NEXT ELEMENT OF THE VECTORS   *
C     * ABOUT TO BE USED.                                                *
C     *********************************************************************
C
      J = J + 1
C
C     *********************************************************************
C     * IF ALL SETS OF BOUNDS AND ASSOCIATED FREQUENCIES HAVE NOT BEEN   *
C     * PROCESSED, BRANCH BACK TO GET ANOTHER;  OTHERWISE CONTINUE TO    *
C     * THE NEXT INSTRUCTION.                                            *
C     *********************************************************************
C
      IF (J.LE.M) GO TO 5
C
C     *********************************************************************
C     * STEP 3(D).  CALCULATE THE ARITHMETIC MEAN (AVG) OF THE FREQUENCY *
C     * TABLE.  THIS IS DONE BY DIVIDING THE SUM OF THE PRODUCTS (XF) OF *
C     * MIDPOINTS AND ASSOCIATED FREQUENCIES BY THE SUM OF THE FREQUEN-  *
C     * CIES (SUMF) IN THE TABLE.                                        *
C     *********************************************************************
C
      AVG = XF / SUMF
C
C     *********************************************************************
C     * STEP 4.  CALCULATE THE NUMERATOR IN THE FORMULA (SUMDXF).        *
C     * INITIALIZE THE ACCUMULATOR OF THE SUM OF THE PRODUCTS OF ABSOL-  *
C     * UTE DEVIATIONS AND FREQUENCIES (SUMDXF) BY SETTING IT EQUAL TO   *
C     * ZERO.  NOTE THAT THIS SUM BECOMES THE NUMERATOR IN THE FORMULA.  *
C     *********************************************************************
C
      SUMDXF = 0.0
C
C     *********************************************************************
C     * REINITIALIZE THE COUNTER AND SUBSCRIPT (J) BY SETTING IT EQUAL   *
C     * TO ONE.  ITS USE HERE WILL BE THE SAME AS ABOVE.                 *
C     *********************************************************************
C
      J = 1
C
C     *********************************************************************
C     * STEP 4(A).  DETERMINE THE ABSOLUTE DEVIATION (ABSDEV) OF THE     *
C     * MIDPOINT FROM THE ARITHMETIC MEAN (AVG) BY SUBTRACTING THE MEAN  *
C     * (AVG) FROM THE MIDPOINT;  NOTE THAT J INDICATES THE NUMBER OF    *
C     * THE ELEMENT CONTAINING THE NEXT MIDPOINT TO BE PROCESSED.  THE   *
C     * ABSOLUTE VALUE CAN BE OBTAINED BY USING THE ABSOLUTE VALUE FUNC- *
C     * TION CALLED ABS.                                                 *
C     *********************************************************************
C
   10 ABSDEV = ABS(X(J) - AVG)
C
C     *********************************************************************
C     * STEP 4(B).  MULTIPLY THE ABSOLUTE DEVIATION (ABSDEV) BY ITS      *
C     * ASSOCIATED FREQUENCY AND STORE THE RESULT IN DEVXF.  NOTE THAT   *
C     * ABSDEV IS THE ABSOLUTE DEVIATION OF THE JTH MIDPOINT, AND THAT   *
C     * ITS ASSOCIATED FREQUENCY IS TO BE FOUND IN THE JTH ELEMENT OF    *
C     * THE VECTOR F, INDICATED BY F(J).                                 *
```

```
C     ********************************************************************
C
      DEVXF = ABSDEV * F(J)
C
C     ********************************************************************
C     * STEP 4(C).  ACCUMULATE THE SUM OF THE ABSOLUTE DEVIATIONS TIMES *
C     * FREQUENCIES (SUMDXF) BY ADDING THE CURRENT VALUE OF THE ABSOLUTE*
C     * DEVIATION TIMES FREQUENCY (DEVXF), OBTAINED IN STEP 4(B), TO THE*
C     * PREVIOUS SUM (SUMDXF).  NOTE THAT THIS SUM BECOMES THE NUMERATOR*
C     * OF THE FORMULA.                                                 *
C     ********************************************************************
C
      SUMDXF = SUMDXF + DEVXF
C
C     ********************************************************************
C     * ADD ONE TO THE COUNTER AND SUBSCRIPT (J) TO INDICATE THE NUMBER *
C     * OF THE NEXT MIDPOINT AND ASSOCIATED FREQUENCY TO BE PROCESSED.  *
C     ********************************************************************
C
      J = J + 1
C
C     ********************************************************************
C     * IF ALL MIDPOINTS AND FREQUENCIES HAVE NOT BEEN PROCESSED, BRANCH*
C     * BACK TO GET ANOTHER; OTHERWISE, CONTINUE TO THE NEXT INSTRUC-   *
C     * TION.                                                           *
C     ********************************************************************
C
      IF (J.LE.M) GO TO 10
C
C     ********************************************************************
C     * STEP 5.  CALCULATE THE MEAN DEVIATION OF THE FREQUENCY TABLE    *
C     * (DEVMF) BY DIVIDING THE SUM OF THE ABSOLUTE DEVIATIONS TIMES    *
C     * FREQUENCIES (SUMDXF) BY THE SUM OF THE FREQUENCIES (SUMF).      *
C     ********************************************************************
C
      DEVMF = SUMDXF / SUMF
C
C     ********************************************************************
C     * PRINT OUT THE MEAN DEVIATION (DEVMF) OF THE FREQUENCY TABLE.    *
C     ********************************************************************
C
      WRITE (6, 600) DEVMF
  600 FORMAT (1X,45HTHE MEAN DEVIATION OF THE FREQUENCY TABLE IS ,
     *  F10.3)
C
C     ********************************************************************
C     * STOP THE PROGRAM.                                               *
C     ********************************************************************
C
      STOP
C
C     ********************************************************************
C     * INDICATE THE END OF THE SOURCE PROGRAM.                         *
C     ********************************************************************
C
      END
```

D. Example Problems

Example 1   Given the following grouped data:

CB	f
15–25	3
25–35	14
35–45	18
45–55	12
55–65	10
65–75	5

find the mean deviation of the above data.

Input Data

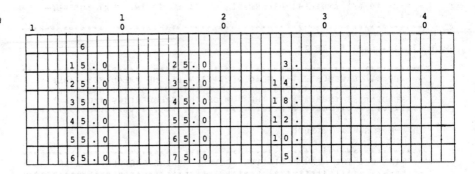

Solution   THE MEAN DEVIATION OF THE FREQUENCY TABLE IS 11.368

Example 2   Given the following grouped data:

CB	f
20–30	2
30–40	5
40–50	8
50–60	3
60–70	2
70–80	5
80–90	2
90–100	1

find the mean deviation of the above data.

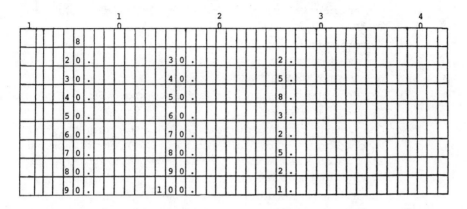

Input Data

THE MEAN DEVIATION OF THE FREQUENCY TABLE IS 16.378   *Solution*

☆ Calculation of the Variance for Grouped Data

A. Statement of Problem

Given a frequency table having class midpoints $X_1, X_2, \ldots, X_M$; respective frequencies $F_1, F_2, \ldots, F_M$, then the variance is defined as:

$$\sigma^2 = \frac{\sum\limits_{i=1}^{M} (X_i - m)^2 F_i}{N}$$

where m is the mean of the data, and N is the sum of the frequencies.

Write a program to find the variance of a frequency table for grouped data.

B. Algorithm

1. Determine the number (M) of lower and upper class bounds (XLOWER, XUPPER) and associated frequencies (F) of the frequency table for which the variance is desired.

2. Insure that the number (M) of lower and upper class bounds (XLOWER, XUPPER) and associated frequencies (F) is 1 or more.

3. Calculate the arithmetic mean (AVG) of the frequency table.

    a. Calculate the midpoint of each class (X) by summing the upper and lower bounds (XUPPER, XLOWER), and dividing by 2.0.

    b. Accumulate the sum of the products (XF) of class midpoints (X) and their associated frequencies (F) for all classes and frequencies in the table.

    c. Accumulate the sum of the frequencies (SUMF) for all classes in the table.

    d. Calculate the arithmetic mean (AVG) of the frequency table by dividing the sum of the products (XF) of midpoints and associated frequencies by the sum of the frequencies (SUMF) in the table.

4. Calculate the numerator in the formula (DEV2XF) by summing the products of the squared deviations of the class midpoints (X) from the mean (AVG) times the associated frequencies (F) for all classes and frequencies in the table.

    a. Determine the deviation (DEV) of the midpoint (X) from the mean (AVG) by subtracting the mean (AVG) from the midpoint (X).

    b. Determine the squared deviation (DEV2) of the midpoint (X) from the mean (AVG) by squaring the deviation (DEV) obtained in Step 4a.

    c. Accumulate the sum of the squared deviations times associated frequencies (DEV2XF) by multiplying the squared deviation (DEV2) by its associated frequency (F) and summing the results of this operation for all classes and frequencies in the table. This sum (DEV2XF) becomes the numerator in the formula.

5. Calculate the variance of the frequency table by dividing the sum of the squared deviations times frequencies (DEV2XF), obtained in Step 4c, by the sum of the frequencies (SUMF), obtained in Step 3c.

C. General Program

```
C
C
C     ***********************************************************************
C     * THIS PROGRAM COMPUTES THE VARIANCE OF A FREQUENCY TABLE.            *
C     ***********************************************************************
C
C
C     ***********************************************************************
C     * INFORM THE COMPUTER THAT X AND F WILL BE VECTORS OF DATA POINTS *
C     * WITH A MAXIMUM OF 300 ELEMENTS.  THESE VECTORS WILL BE USED TO   *
```

```
C     * STORE THE CLASS MIDPOINTS AND ASSOCIATED FREQUENCIES OF THE TA- *
C     * BLE FOR FUTURE REFERENCE.  NOTE THAT THE CHOICE OF 300 AS A LIM-*
C     * IT IS ARBITRARY.                                                *
C     ******************************************************************
C
      DIMENSION X(300), F(300)
C
C     ******************************************************************
C     * STEP 1.  DETERMINE THE NUMBER (M) OF LOWER AND UPPER CLASS      *
C     * BOUNDS (XLOWER, XUPPER) AND ASSOCIATED FREQUENCIES (F) IN THE   *
C     * TABLE.                                                          *
C     ******************************************************************
C
      READ (5, 500) M
  500 FORMAT (I6)
C
C     ******************************************************************
C     * STEP 2.  TEST THE NUMBER (M) OF LOWER AND UPPER CLASS BOUNDS    *
C     * (XLOWER, XUPPER) AND ASSOCIATED FREQUENCIES (F) TO INSURE IT IS *
C     * 1 OR MORE; IF IT IS NOT, THE PROGRAM STOPS.                     *
C     ******************************************************************
C
      IF (M .LT. 1) STOP
C
C     ******************************************************************
C     * TEST THE NUMBER (M) OF LOWER AND UPPER CLASS BOUNDS (XLOWER,    *
C     * XUPPER) AND ASSOCIATED FREQUENCIES (F) TO INSURE IT IS 300 OR   *
C     * LESS; IF IT IS NOT, THE PROGRAM STOPS.  NOTE THAT 300 IS AN AR- *
C     * BITRARY RESTRICTION OF THE PROGRAM, BUT IT MUST BE ENFORCED     *
C     * SINCE THE VECTORS X AND F CAN CONTAIN A MAXIMUM OF 300 ELEMENTS.*
C     ******************************************************************
C
      IF (M .GT. 300) STOP
C
C     ******************************************************************
C     * STEP 3.  CALCULATE THE ARITHMETIC MEAN (AVG) OF THE FREQUENCY   *
C     * TABLE.                                                          *
C     ******************************************************************
C
C
C     ******************************************************************
C     * INITIALIZE THE COUNTER AND SUBSCRIPT (J) BY SETTING IT EQUAL TO *
C     * ONE.  THIS WILL BE USED TO INDICATE THE NUMBER OF THE SET OF    *
C     * BOUNDS AND ASSOCIATED FREQUENCIES (XLOWER, XUPPER, F) ABOUT TO  *
C     * BE PROCESSED.  IT WILL ALSO BE USED AS A SUBSCRIPT TO REFERENCE *
C     * ANY PARTICULAR ELEMENT OF EITHER OF THE VECTORS IN THE PROGRAM. *
C     ******************************************************************
C
      J = 1
C
C     ******************************************************************
C     * INITIALIZE THE ACCUMULATOR OF THE PRODUCTS (XF) OF CLASS MID-   *
C     * POINTS AND ASSOCIATED FREQUENCIES BY SETTING IT EQUAL TO ZERO.  *
C     ******************************************************************
C
      XF = 0.0
C
C     ******************************************************************
C     * INITIALIZE THE ACCUMULATOR OF THE SUM OF THE FREQUENCIES (SUMF) *
C     * BY SETTING IT EQUAL TO ZERO.                                    *
C     ******************************************************************
C
      SUMF = 0.0
```

```
C
C     **********************************************************************
C     * GET THE NEXT SET OF BOUNDS AND ASSOCIATED FREQUENCY (XLOWER,      *
C     * XUPPER, F).  AT THIS POINT, J INDICATES THE NUMBER OF THE SET OF* 
C     * BOUNDS AND ASSOCIATED FREQUENCY ABOUT TO BE READ.  IT WILL BE     *
C     * FOUND CONVENIENT TO READ THE JTH FREQUENCY INTO THE JTH ELEMENT   *
C     * OF THE VECTOR F.  NOTE THAT THE JTH ELEMENT OF VECTOR F IS INDI-* 
C     * CATED BY F(J).                                                    *
C     **********************************************************************
C
  5      READ (5, 501) XLOWER, XUPPER, F(J)
  501    FORMAT (3F10.3)
C
C     **********************************************************************
C     * STEP 3(A).  CALCULATE THE MIDPOINT OF THIS CLASS BY SUMMING ITS   *
C     * LOWER AND UPPER BOUNDS AND DIVIDING THE RESULT BY 2.  AGAIN IT     *
C     * WILL BE FOUND CONVENIENT TO STORE THE MIDPOINT OF THE JTH CLASS   *
C     * IN THE JTH ELEMENT OF THE VECTOR X.                               *
C     **********************************************************************
C
         X(J) = (XLOWER + XUPPER) / 2.0
C
C     **********************************************************************
C     * STEP 3(B).  ACCUMULATE THE SUM OF THE PRODUCTS (XF) OF THE CLASS* 
C     * MIDPOINTS AND ASSOCIATED FREQUENCIES BY MULTIPLYING THE MIDPOINT* 
C     * BY THE FREQUENCY AND ADDING THE RESULT OF THIS TO THE PREVIOUS    *
C     * SUM OF THE PRODUCTS (XF).  NOTE THAT J INDICATES THE NUMBER OF    *
C     * THE ELEMENT OF THE VECTORS WHERE THE CURRENT MIDPOINT AND ITS     *
C     * ASSOCIATED FREQUENCY ARE STORED.  THUS X(J) INDICATES THE CUR-    *
C     * RENT MIDPOINT AND F(J) INDICATES THE ASSOCIATED FREQUENCY.        *
C     **********************************************************************
C
         XF = XF + (X(J) * F(J))
C
C     **********************************************************************
C     * STEP 3(C).  ACCUMULATE THE SUM OF THE FREQUENCIES (SUMF) BY ADD-* 
C     * ING THE VALUE OF THE CURRENT FREQUENCY, INDICATED BY F(J), TO     *
C     * THE PREVIOUS VALUE OF THE SUM OF THE FREQUENCIES (SUMF).          *
C     **********************************************************************
C
         SUMF = SUMF + F(J)
C
C     **********************************************************************
C     * ADD ONE TO THE COUNTER AND SUBSCRIPT (J) TO INDICATE THE NUMBER   *
C     * OF THE NEXT SET OF BOUNDS AND ASSOCIATED FREQUENCY ABOUT TO BE    *
C     * PROCESSED AND ALSO TO INDICATE THE NEXT ELEMENT OF THE VECTORS    *
C     * ABOUT TO BE USED.                                                 *
C     **********************************************************************
C
         J = J + 1
C
C     **********************************************************************
C     * IF ALL SETS OF BOUNDS AND ASSOCIATED FREQUENCIES HAVE NOT BEEN    *
C     * PROCESSED, BRANCH BACK TO GET ANOTHER;  OTHERWISE CONTINUE TO     *
C     * THE NEXT INSTRUCTION.                                             *
C     **********************************************************************
C
         IF (J .LE. M) GO TO 5
C
C     **********************************************************************
C     * STEP 3(D).  CALCULATE THE ARITHMETIC MEAN (AVG) OF THE FREQUENCY* 
C     * TABLE.  THIS IS DONE BY DIVIDING THE SUM OF THE PRODUCTS (XF) OF* 
C     * MIDPOINTS AND ASSOCIATED FREQUENCIES BY THE SUM OF THE FREQUEN-   *
C     * CIES (SUMF) IN THE TABLE.                                         *
```

```
C     ******************************************************************
C
C         AVG = XF / SUMF
C
C     ******************************************************************
C     * STEP 4.  CALCULATE THE NUMERATOR OF THE FORMULA (DEV2XF).      *
C     ******************************************************************
C
C
C     ******************************************************************
C     * INITIALIZE THE ACCUMULATOR OF THE SUM OF THE PRODUCTS OF ABSOL-*
C     * UTE DEVIATIONS AND FREQUENCIES (SUMDXF) BY SETTING IT EQUAL TO *
C     * ZERO.  NOTE THAT THIS SUM BECOMES THE NUMERATOR IN THE FORMULA.*
C     ******************************************************************
C
      DEV2XF = 0.0
C
C     ******************************************************************
C     * REINITIALIZE THE COUNTER AND SUBSCRIPT (J) BY SETTING IT EQUAL *
C     * TO ONE.  ITS USE HERE WILL BE THE SAME AS ABOVE.              *
C     ******************************************************************
C
      J = 1
C
C     ******************************************************************
C     * STEP 4(A).  CALCULATE THE DEVIATION (DEV) OF THE MIDPOINT (X)  *
C     * FROM THE MEAN (AVG) BY SUBTRACTING THE MEAN (AVG) FROM THE MID-*
C     * POINT (X); NOTE THAT J INDICATES THE NUMBER OF THE ELEMENT CON-*
C     * TAINING THE NEXT MIDPOINT TO BE PROCESSED.                     *
C     ******************************************************************
C
   10 DEV = (X(J) - AVG)
C
C     ******************************************************************
C     * STEP 4(B).  SQUARE THE DEVIATION (DEV) AND STORE THE RESULT IN *
C     * DEV2.  NOTE THAT THE DOUBLE ASTERISK (**) INDICATES EXPONENTIA-*
C     * TION; THE 2 INDICATES THE POWER TO WHICH DEV IS TO BE RAISED.  *
C     ******************************************************************
C
      DEV2 = DEV ** 2
C
C     ******************************************************************
C     * STEP 4(C).  MULTIPLY THE SQUARED DEVIATION (DEV2) BY ITS ASSOCI-*
C     * ATED FREQUENCY AND ACCUMULATE THE SUM OF THE RESULTS.  DEV2 IS *
C     * THE SQUARED DEVIATION OF THE JTH MIDPOINT, AND ITS ASSOCIATED  *
C     * FREQUENCY IS TO BE FOUND IN THE JTH ELEMENT OF THE VECTOR F, IN-*
C     * DICATED BY F(J).                                               *
C     ******************************************************************
C
      DEV2XF = DEV2XF + (DEV2 * F(J))
C
C     ******************************************************************
C     * ADD ONE TO THE COUNTER AND SUBSCRIPT (J) TO INDICATE THE NUMBER *
C     * OF THE NEXT MIDPOINT AND ASSOCIATED FREQUENCY TO BE PROCESSED. *
C     ******************************************************************
C
      J = J + 1
C
C     ******************************************************************
C     * IF ALL MIDPOINTS AND FREQUENCIES HAVE NOT BEEN PROCESSED, BRANCH*
C     * BACK TO GET ANOTHER; OTHERWISE, CONTINUE TO THE NEXT INSTRUC-  *
C     * TION.                                                          *
C     ******************************************************************
C
      IF (J .LE. M) GO TO 10
```

```
C
C      *******************************************************************
C      * STEP 5.  CALCULATE THE VARIANCE OF THE FREQUENCY TABLE BY DIVID-*
C      * ING THE NUMERATOR IN THE FORMULA (DEV2XF) BY THE SUM OF THE FRE-*
C      * QUENCIES (SUMF).                                                *
C      *******************************************************************
C
       VAR = DEV2XF / SUMF
C
C      *******************************************************************
C      * PRINT OUT THE VARIANCE OF THE FREQUENCY TABLE (VAR).            *
C      *******************************************************************
C
       WRITE (6, 600) VAR
  600  FORMAT (1X,39HTHE VARIANCE OF THE FREQUENCY TABLE IS , F10.3)
C
C      *******************************************************************
C      * STOP THE PROGRAM.                                              *
C      *******************************************************************
C
       STOP
C
C      *******************************************************************
C      * INDICATE THE END OF THE SOURCE PROGRAM.                        *
C      *******************************************************************
C
       END
```

D. Example Problems

Example 1   Given the following grouped data:

CB	f
15–25	3
25–35	14
35–45	18
45–55	12
55–65	10
65–75	5

find the variance of the above data.

Input Data

```
    6
 15.0        25.0        3.
 25.0        35.0       14.
 35.0        45.0       18.
 45.0        55.0       12.
 55.0        65.0       10.
 65.0        75.0        5.
```

THE VARIANCE OF THE FREQUENCY TABLE IS 179.422

Solution

Given the following grouped data:

Example 2

CB	f
20–30	2
30–40	5
40–50	8
50–60	3
60–70	2
70–80	5
80–90	2
90–100	1

find the variance of the above data.

Input Data

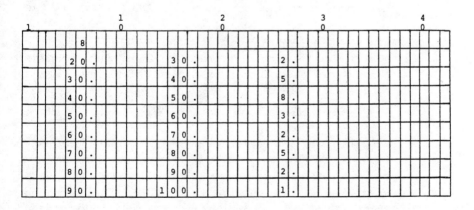

```
    8
 20.         30.         2.
 30.         40.         5.
 40.         50.         8.
 50.         60.         3.
 60.         70.         2.
 70.         80.         5.
 80.         90.         2.
 90.        100.         1.
```

THE VARIANCE OF THE FREQUENCY TABLE IS 363.775

Solution

☆ **Calculation of the Standard Deviation of Grouped Data**

A. Statement of Problem

Given a frequency table with class midpoints X_1, X_2,..., X_M, and respective frequencies F_1, F_2,..., F_M, then we can define the standard deviation as:

$$\sigma = \sqrt{\frac{\displaystyle\sum_{i=1}^{M} (X_i - m)^2 F_i}{N}}$$

where M = number of classes within the frequency table.
N = sum of the frequencies.
m = the mean of the variates.

Write a computer program to calculate the standard deviation of a frequency table.

B. Algorithm

1. Determine the number (M) of lower and upper class bounds (XLOWER, XUPPER) and associated frequencies (F) of the frequency table for which the standard deviation is desired.

2. Insure that the number (M) of lower and upper class bounds (XLOWER, XUPPER) and associated frequencies (F) is 1 or more.

3. Calculate the arithmetic mean (AVG) of the frequency table.

   a. Calculate the midpoint of each class (X) by summing the upper and lower bounds (XUPPER, XLOWER) and dividing by 2.0.

   b. Accumulate the sum of the products (XF) of class midpoints (X) and their associated frequencies (F) for all classes and frequencies in the table.

   c. Accumulate the sum of the frequencies (SUMF) for all classes in the table.

   d. Calculate the arithmetic mean (AVG) of the frequency table by dividing the sum of the products (XF) of midpoints and associated frequencies (SUMF) in the table.

4. Calculate the numerator in the formula (DEV2XF) by summing the products of the squared deviations of class midpoints (X) from the mean (AVG) by subtracting the mean (AVG) times

the associated frequencies (F) for all classes and frequencies in the table.

   a. Determine the squared deviation (DEV2) of the midpoint (X) from the mean (AVG) from the midpoint (X), squaring the result.

   b. Accumulate the sum of the squared deviations times associated frequencies (DEV2XF) by multiplying the squared deviation (DEV2) by its associated frequency (F) and summing the results of this operation for all classes and frequencies in the table. This sum (DEV2XF) becomes the numerator in the formula.

5. Calculate the standard deviation of the frequency table (SD) by dividing the sum of the squared deviations times frequencies (DEV2XF) obtained in Step 4b and then extracting the square root of the result.

C. General Program

```
C
C
C     ******************************************************************
C     * THIS PROGRAM COMPUTES THE STANDARD DEVIATION OF A FREQUENCY    *
C     * TABLE.                                                         *
C     ******************************************************************
C
C
C
C     ******************************************************************
C     * INFORM THE COMPUTER THAT X AND F WILL BE VECTORS OF DATA POINTS *
C     * WITH A MAXIMUM OF 300 ELEMENTS.  THESE VECTORS WILL BE USED TO  *
C     * STORE THE CLASS MIDPOINTS AND ASSOCIATED FREQUENCIES OF THE TA- *
C     * BLE FOR FUTURE REFERENCE.  NOTE THAT THE CHOICE OF 300 AS A LIM-*
C     * IT IS ARBITRARY.                                               *
C     ******************************************************************
C
      DIMENSION X(300), F(300)
C
C     ******************************************************************
C     * STEP 1.  DETERMINE THE NUMBER (M) OF LOWER AND UPPER CLASS      *
C     * BOUNDS (XLOWER, XUPPER) AND ASSOCIATED FREQUENCIES (F) IN THE   *
C     * TABLE.                                                         *
C     ******************************************************************
C
      READ (5, 500) M
  500 FORMAT (I6)
C
C     ******************************************************************
C     * STEP 2.  TEST THE NUMBER (M) OF LOWER AND UPPER CLASS BOUNDS    *
C     * (XLOWER, XUPPER) AND ASSOCIATED FREQUENCIES (F) TO INSURE IT IS *
C     * 1 OR MORE; IF IT IS NOT, THE PROGRAM STOPS.                    *
C     ******************************************************************
C
      IF (M .LT. 1) STOP
C
C     ******************************************************************
C     * TEST THE NUMBER (M) OF LOWER AND UPPER CLASS BOUNDS (XLOWER,    *
C     * XUPPER) AND ASSOCIATED FREQUENCIES (F) TO INSURE IT IS 300 OR   *
```

```
C     * LESS; IF IT IS NOT, THE PROGRAM STOPS.  NOTE THAT 300 IS AN AR- *
C     * BITRARY RESTRICTION OF THE PROGRAM, BUT IT MUST BE ENFORCED      *
C     * SINCE THE VECTORS X AND F CAN CONTAIN A MAXIMUM OF 300 ELEMENTS.*
C     ****************************************************************
C
      IF (M .GT. 300) STOP
C
C     ****************************************************************
C     * STEP 3.  CALCULATE THE ARITHMETIC MEAN (AVG) OF THE FREQUENCY  *
C     * TABLE.                                                         *
C     ****************************************************************
C
C
C
C     ****************************************************************
C     * INITIALIZE THE COUNTER AND SUBSCRIPT (J) BY SETTING IT EQUAL TO *
C     * ONE.  THIS WILL BE USED TO INDICATE THE NUMBER OF THE SET OF    *
C     * BOUNDS AND ASSOCIATED FREQUENCIES (XLOWER, XUPPER, F) ABOUT TO  *
C     * BE PROCESSED.  IT WILL ALSO BE USED AS A SUBSCRIPT TO REFERENCE *
C     * ANY PARTICULAR ELEMENT OF EITHER OF THE VECTORS IN THE PROGRAM. *
C     ****************************************************************
C
      J = 1
C
C     ****************************************************************
C     * INITIALIZE THE ACCUMULATOR OF THE PRODUCTS (XF) OF CLASS MID-   *
C     * POINTS AND ASSOCIATED FREQUENCIES BY SETTING IT EQUAL TO ZERO.  *
C     ****************************************************************
C
      XF = 0.0
C
C     ****************************************************************
C     * INITIALIZE THE ACCUMULATOR OF THE SUM OF THE FREQUENCIES (SUMF) *
C     * BY SETTING IT EQUAL TO ZERO.                                    *
C     ****************************************************************
C
      SUMF = 0.0
C
C     ****************************************************************
C     * GET THE NEXT SET OF BOUNDS AND ASSOCIATED FREQUENCY (XLOWER,    *
C     * XUPPER, F).  AT THIS POINT, J INDICATES THE NUMBER OF THE SET OF*
C     * BOUNDS AND ASSOCIATED FREQUENCY ABOUT TO BE READ.  IT WILL BE   *
C     * FOUND CONVENIENT TO READ THE JTH FREQUENCY INTO THE JTH ELEMENT *
C     * OF THE VECTOR F.  NOTE THAT THE JTH ELEMENT OF VECTOR F IS INDI-*
C     * CATED BY F(J).                                                  *
C     ****************************************************************
C
    5 READ (5, 501) XLOWER, XUPPER, F(J)
  501 FORMAT (3F10.3)
C
C     ****************************************************************
C     * STEP 3(A).  CALCULATE THE MIDPOINT OF THIS CLASS BY SUMMING ITS *
C     * LOWER AND UPPER BOUNDS AND DIVIDING THE RESULT BY 2.  AGAIN IT  *
C     * WILL BE FOUND CONVENIENT TO STORE THE MIDPOINT OF THE JTH CLASS *
C     * IN THE JTH ELEMENT OF THE VECTOR X.                             *
C     ****************************************************************
C
      X(J) = (XLOWER + XUPPER) / 2.0
C
C     ****************************************************************
C     * STEP 3(B).  ACCUMULATE THE SUM OF THE PRODUCTS (XF) OF THE CLASS*
C     * MIDPOINTS AND ASSOCIATED FREQUENCIES BY MULTIPLYING THE MIDPOINT*
C     * BY THE FREQUENCY AND ADDING THE RESULT OF THIS TO THE PREVIOUS  *
C     * SUM OF THE PRODUCTS (XF).  NOTE THAT J INDICATES THE NUMBER OF  *
C     * THE ELEMENT OF THE VECTORS WHERE THE CURRENT MIDPOINT AND ITS   *
```

```
C     * ASSOCIATED FREQUENCY ARE STORED.  THUS X(J) INDICATES THE CUR-  *
C     * RENT MIDPOINT AND F(J) INDICATES THE ASSOCIATED FREQUENCY.       *
C     ********************************************************************
C
      XF = XF + (X(J) * F(J))
C
C     ********************************************************************
C     * STEP 3(C).  ACCUMULATE THE SUM OF THE FREQUENCIES (SUMF) BY ADD-*
C     * ING THE VALUE OF THE CURRENT FREQUENCY, INDICATED BY F(J), TO    *
C     * THE PREVIOUS VALUE OF THE SUM OF THE FREQUENCIES (SUMF).         *
C     ********************************************************************
C
      SUMF = SUMF + F(J)
C
C     ********************************************************************
C     * ADD ONE TO THE COUNTER AND SUBSCRIPT (J) TO INDICATE THE NUMBER *
C     * OF THE NEXT SET OF BOUNDS AND ASSOCIATED FREQUENCY ABOUT TO BE   *
C     * PROCESSED AND ALSO TO INDICATE THE NEXT ELEMENT OF THE VECTORS   *
C     * ABOUT TO BE USED.                                                *
C     ********************************************************************
C
      J = J + 1
C
C     ********************************************************************
C     * IF ALL SETS OF BOUNDS AND ASSOCIATED FREQUENCIES HAVE NOT BEEN  *
C     * PROCESSED, BRANCH BACK TO GET ANOTHER;  OTHERWISE CONTINUE TO    *
C     * THE NEXT INSTRUCTION.                                            *
C     ********************************************************************
C
      IF (J .LE. M) GO TO 5
C
C     ********************************************************************
C     * STEP 3(D).  CALCULATE THE ARITHMETIC MEAN (AVG) OF THE FREQUENCY*
C     * TABLE.  THIS IS DONE BY DIVIDING THE SUM OF THE PRODUCTS (XF) OF*
C     * MIDPOINTS AND ASSOCIATED FREQUENCIES BY THE SUM OF THE FREQUEN- *
C     * CIES (SUMF) IN THE TABLE.                                        *
C     ********************************************************************
C
      AVG = XF / SUMF
C
C     ********************************************************************
C     * STEP 4.  CALCULATE THE NUMERATOR OF THE FORMULA (DEV2XF).        *
C     ********************************************************************
C
C
C     ********************************************************************
C     * INITIALIZE THE ACCUMULATOR OF THE PRODUCT OF THE SQUARED DEVIA- *
C     * TIONS TIMES ASSOCIATED FREQUENCIES (DEV2XF) BY SETTING IT EQUAL  *
C     * TO ZERO.                                                         *
C     ********************************************************************
C
      DEV2XF = 0.0
C
C     ********************************************************************
C     * REINITIALIZE THE COUNTER AND SUBSCRIPT (J) BY SETTING IT EQUAL   *
C     * TO ONE.  ITS USE HERE WILL BE THE SAME AS ABOVE.                 *
C     ********************************************************************
C
      J = 1
C
C     ********************************************************************
C     * STEP 4(A).  CALCULATE THE SQUARED DEVIATION OF THE MIDPOINT (X) *
C     * FROM THE MEAN (AVG) BY SUBTRACTING THE MEAN (AVG) FROM THE MID- *
C     * POINT (X) AND SQUARING THE RESULT.  NOTE THAT J INDICATES THE    *
```

63

```
C     * NUMBER OF THE ELEMENT CONTAINING THE NEXT MIDPOINT TO BE PRO-   *
C     * CESSED.  NOTE ALSO THAT THE DOUBLE ASTERISK (**) INDICATES EX-  *
C     * PONENTIATION AND THE 2 INDICATES THE POWER TO WHICH THE RESULT  *
C     * OF THE SUBTRACTION IS TO BE RAISED.                             *
C     *****************************************************************
C
  10    DEV2 = (X(J) - AVG) ** 2
C
C     *****************************************************************
C     * STEP 4(B).  MULTIPLY THE SQUARED DEVIATION (DEV2) BY ITS ASSOCI-*
C     * ATED FREQUENCY AND ACCUMULATE THE SUM OF THE RESULTS.  DEV2 IS  *
C     * THE SQUARED DEVIATION OF THE JTH MIDPOINT, AND ITS ASSOCIATED   *
C     * FREQUENCY IS TO BE FOUND IN THE JTH ELEMENT OF THE VECTOR F, IN-*
C     * DICATED BY F(J).                                                *
C     *****************************************************************
C
        DEV2XF = DEV2XF + (DEV2 * F(J))
C
C     *****************************************************************
C     * ADD ONE TO THE COUNTER AND SUBSCRIPT (J) TO INDICATE THE NUMBER *
C     * OF THE NEXT MIDPOINT AND ASSOCIATED FREQUENCY TO BE PROCESSED.  *
C     *****************************************************************
C
        J = J + 1
C
C     *****************************************************************
C     * IF ALL MIDPOINTS AND FREQUENCIES HAVE NOT BEEN PROCESSED, BRANCH*
C     * BACK TO GET ANOTHER; OTHERWISE, CONTINUE TO THE NEXT INSTRUC-   *
C     * TION.                                                           *
C     *****************************************************************
C
        IF (J .LE. M) GO TO 10
C
C     *****************************************************************
C     * STEP 5.  CALCULATE THE STANDARD DEVIATION OF THE FREQUENCY TABLE*
C     * BY DIVIDING THE NUMERATOR OF THE FORMULA (DEV2XF) BY THE SUM OF *
C     * THE FREQUENCIES (SUMF) AND EXTRACTING THE SQUARE ROOT OF THIS   *
C     * RESULT.  THE SQUARE ROOT MAY BE OBTAINED BY RAISING THE RESULT  *
C     * OF THE DIVISION TO THE ONE-HALF (0.5) POWER.                    *
C     *****************************************************************
C
        SD = (DEV2XF / SUMF) ** 0.5
C
C     *****************************************************************
C     * PRINT OUT THE STANDARD DEVIATION OF THE FREQUENCY TABLE (SD).   *
C     *****************************************************************
C
        WRITE (6, 600) SD
  600   FORMAT (1X,49HTHE STANDARD DEVIATION OF THE FREQUENCY TABLE IS ,
       *  F10.3)
C
C     *****************************************************************
C     * STOP THE PROGRAM.                                              *
C     *****************************************************************
C
        STOP
C
C     *****************************************************************
C     * INDICATE THE END OF THE SOURCE PROGRAM.                        *
C     *****************************************************************
C
        END
```

D. Example Problems

Given the following grouped data:

Example 1

CB	f
15–25	3
25–35	14
35–45	18
45–55	12
55–65	10
65–75	5

find the standard deviation of the above data.

Input Data

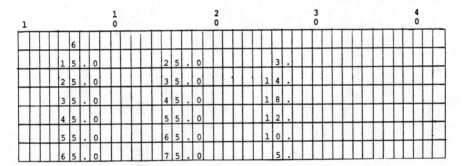

THE STANDARD DEVIATION OF THE FREQUENCY TABLE IS 13.395   *Solution*

Given the following grouped data:

Example 2

CB	f
20–30	2
30–40	5
40–50	8
50–60	3
60–70	2
70–80	5
80–90	2
90–100	1

find the standard deviation of the above data.

Input Data

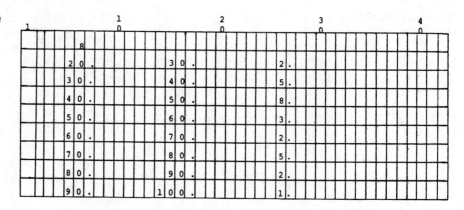

Solution   THE STANDARD DEVIATION OF THE FREQUENCY TABLE IS 19.073

1. Given the following distribution of weights for a varsity football team:

Class boundary	Frequency
120–140	5
140–160	8
160–180	12
180–200	13
200–220	4
220–240	2

find:

a. The mean of the above distribution.

b. The mean deviation.

c. The variance.

d. The standard deviation.

2. Given the following distribution of battery lives as measured in years:

Class boundary	Frequency
0.5–1.5	2
1.5–2.5	3
2.5–3.5	18
3.5–4.5	10
4.5–5.5	6
5.5–6.5	1

find:

a. The mean of the above distribution.

b. The mean deviation.

c. The variance.

d. The standard deviation.

☆ **Calculation of a Weighted Mean**

A. Statement of Problem

Given a set of N_1 variates with a mean m_1, and a set of N_2 variates with a mean m_2, using equation

$$m_w = \frac{N_1 m_1 + N_2 m_2 + \ldots + N_K m_K}{N_1 + N_2 + \ldots + N_K}$$

write a computer program for calculating a weighted mean for *k* sets of variables.

B. Algorithm

1. Determine the number (K) of pairs of group sizes and associated means (N, M) for which the weighted mean is desired.

2. Insure that the number (K) of pairs of group sizes and associated means (N, M) is 2 or more.

3. Accumulate the sum of the group sizes (SUMN) and the sum of the products (SUMNXM) of group sizes and associated means for all pairs of group sizes and means.

4. Calculate the weighted mean (WMEAN) by dividing the sum of the products (SUMNXM) of group sizes and associated means by the sum of the group sizes (SUMN).

C. General Program

```
C
C
C     ****************************************************************
C     * THIS PROGRAM COMPUTES A WEIGHTED MEAN FROM A SERIES OF PAIRS OF *
C     * GROUP SIZES AND ASSOCIATED ARITHMETIC MEANS.                 *
C     ****************************************************************
C
C
C
C     ****************************************************************
C     * INFORM THE COMPUTER THAT THE VARIABLE M WILL BE A REAL RATHER *
C     * THAN AN INTEGER VARIABLE.                                    *
C     ****************************************************************
C
      RFAL M
C
C
C     ****************************************************************
C     * STEP 1.  DETERMINE THE NUMBER (K) OF PAIRS OF GROUP SIZES AND *
C     * ASSOCIATED MEANS (N, M) FOR WHICH THE WEIGHTED AVERAGE IS DE- *
C     * SIRED.                                                       *
C     ****************************************************************
C
      READ (5, 500) K
  500 FORMAT (I6)
C
C
C     ****************************************************************
C     * STEP 2.  TEST THE NUMBER (K) OF PAIRS OF GROUP SIZES AND ASSOCI-*
C     * ATED MEANS (N, M) TO INSURE IT IS 2 OR MORE; IF IT IS NOT, THE *
C     * PROGRAM STOPS.                                               *
C     ****************************************************************
C
      IF (K .LT. 2) STOP
C
C
C     ****************************************************************
C     * STEP 3.  ACCUMULATE THE SUM OF THE GROUP SIZES (SUMN) AND THE *
```

```
C      * SUM OF THE PRODUCTS (SUMNXM) OF GROUP SIZES AND ASSOCIATED      *
C      * MEANS.                                                          *
C
C      ******************************************************************
C
C      ******************************************************************
C      * INITIALIZE THE COUNTER (J) BY SETTING IT EQUAL TO ZERO.  THIS   *
C      * WILL BE USED TO COUNT THE NUMBER OF PAIRS OF GROUP SIZES AND    *
C      * ASSOCIATED MEANS (N, M) THAT HAVE BEEN PROCESSED.               *
C      ******************************************************************
C
       J = 0
C
C      ******************************************************************
C      * INITIALIZE THE ACCUMULATORS OF THE SUM OF THE GROUP SIZES (SUMN)*
C      * AND THE SUM OF THE PRODUCTS (SUMNXM) OF GROUP SIZES AND ASSOCIA-*
C      * TED MEANS BY SETTING THEM EQUAL TO ZERO.                        *
C      ******************************************************************
C
       SUMN = 0.0
       SUMNXM = 0.0
C
C      ******************************************************************
C      * GET THE NEXT PAIR OF GROUP SIZES AND ASSOCIATED MEANS (N, M).   *
C      ******************************************************************
C
  100  READ (5, 501) N, M
  501  FORMAT (I6, F6.2)
C
C      ******************************************************************
C      * ACCUMULATE THE SUM OF THE GROUP SIZES (SUMN) BY ADDING THE VALUE*
C      * OF THE CURRENT GROUP SIZE (N) TO THE PREVIOUS VALUE OF THE SUM  *
C      * OF THE GROUP SIZES (SUMN).                                      *
C      ******************************************************************
C
       SUMN = SUMN + N
C
C      ******************************************************************
C      * ACCUMULATE THE SUM OF THE PRODUCTS (SUMNXM) OF GROUP SIZES AND  *
C      * ASSOCIATED MEANS BY MULTIPLYING THE CURRENT GROUP SIZE (N) BY   *
C      * ITS ASSOCIATED MEAN (M) AND ADDING THIS RESULT TO THE PREVIOUS  *
C      * VALUE OF THE SUM OF THE PRODUCTS (SUMNXM).                      *
C      ******************************************************************
C
       SUMNXM = SUMNXM + (N * M)
C
C      ******************************************************************
C      * ADD ONE TO THE COUNTER (J) TO INDICATE THAT ANOTHER PAIR OF     *
C      * GROUP SIZES AND ASSOCIATED MEANS (N, M) HAS BEEN PROCESSED.     *
C      ******************************************************************
C
       J = J + 1
C
C      ******************************************************************
C      * IF ALL PAIRS (N, M) HAVE NOT BEEN PROCESSED, BRANCH BACK TO GET *
C      * ANOTHER; OTHERWISE, CONTINUE TO THE NEXT INSTRUCTION.           *
C      ******************************************************************
C
       IF (J .LT. K) GO TO 100
C
C      ******************************************************************
C      * STEP 4.  CALCULATE THE WEIGHTED MEAN (WMEAN) BY DIVIDING THE SUM*
C      * OF THE PRODUCTS (SUMNXM) OF GROUP SIZES AND MEANS BY THE SUM OF *
C      * THE GROUP SIZES (SUMN).                                         *
```

```
C
C     ***********************************************************************
C
      WMEAN = SUMNXM / SUMN
C
C     ***********************************************************************
C     * PRINT OUT THE WEIGHTED MEAN (WMEAN).                                *
C     ***********************************************************************
C
      WRITE (6, 600) WMEAN
  600 FORMAT (1X,21HTHE WEIGHTED MEAN IS , F10.3)
C
C     ***********************************************************************
C     * STOP THE PROGRAM.                                                   *
C     ***********************************************************************
C
      STOP
C
C     ***********************************************************************
C     * INDICATE THE END OF THE SOURCE PROGRAM.                             *
C     ***********************************************************************
C
      END
```

D. Example Problems

Example 1   A statistics class is divided into two sections, both of which are given the same test. Section I (50 students) has a mean of 68; Section II (42 students) has a mean of 64. Find the weighted mean of the two sections.

Input Data

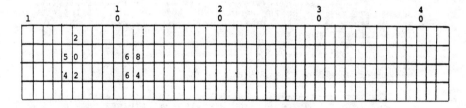

Solution   THE WEIGHTED MEAN IS 66.174

Example 2   A manufacturing company announced its Christmas bonus awards. A total of 100 white collar workers received bonus awards of $6000 each, while a total of 200 blue collar workers received bonus awards of $4000 each. Compute the weighted average of the bonus received by both groups.

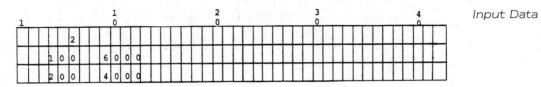

Input Data

THE WEIGHTED MEAN IS 4666.664

Solution

☆ Calculation of a Harmonic Mean

A. Statement of Problem

Given a set of N variates $X_1, X_2, \ldots, X_N$, using the equation

$$m_h = \frac{N}{\dfrac{1}{X_1} + \dfrac{1}{X_2} + \ldots + \dfrac{1}{X_N}}$$

write a computer program which will calculate the harmonic mean of N variates.

B. Algorithm

1. Determine the number (N) of data points (X) for which the harmonic mean is desired.

2. Test the number (N) of data points (X) to insure that it is 1 or more.

3. Calculate the denominator of the formula by accumulating the sum of the reciprocals (SUMREC) of all data points.

4. Calculate the harmonic mean (HMEAN) by dividing the number (N) of data points by the sum of the reciprocals (SUMREC) of the data points.

C. General Program

```
C
C
C      ***********************************************************************
C      * THIS PROGRAM COMPUTES THE HARMONIC MEAN OF A SERIES OF DATA        *
C      * POINTS.                                                            *
C      ***********************************************************************
C
C
C
C      ***********************************************************************
C      * STEP 1.  DETERMINE THE NUMBER (N) OF DATA POINTS (X) IN THE SER-*
C      * IES.                                                               *
C      ***********************************************************************
C
       READ (5, 500) N
  500  FORMAT (I6)
```

```
C
C     ****************************************************************
C     * STEP 2.  TEST THE NUMBER (N) OF DATA POINTS (X) TO INSURE IT IS *
C     * 1 OR MORE.                                                    *
C     ****************************************************************
C
      IF (N .LT. 1) STOP
C
C     ****************************************************************
C     * STEP 3.  DETERMINE THE SUM (SUMREC) OF THE RECIPROCALS OF THE *
C     * DATA POINTS.                                                  *
C     ****************************************************************
C
C
C     ****************************************************************
C     * INITIALIZE THE COUNTER (K) BY SETTING IT EQUAL TO ZERO; THIS  *
C     * WILL BE USED TO COUNT THE NUMBER OF DATA POINTS (X) THAT HAVE  *
C     * BEEN PROCESSED.                                               *
C     ****************************************************************
C
      K = 0
C
C     ****************************************************************
C     * INITIALIZE THE ACCUMULATOR (SUMREC) OF THE RECIPROCALS OF THE *
C     * DATA POINTS (X) BY SETTING IT EQUAL TO ZERO.                  *
C     ****************************************************************
C
      SUMREC = 0.0
C
C     ****************************************************************
C     * GET THE NEXT DATA POINT (X).                                  *
C     ****************************************************************
C
    1 READ (5, 501) X
  501 FORMAT (F10.3)
C
C     ****************************************************************
C     * ACCUMULATE THE SUM (SUMREC) OF THE RECIPROCALS BY DIVIDING THE *
C     * VALUE OF THE CURRENT DATA POINT (X) INTO ONE AND ADDING THE RE- *
C     * SULT TO THE PREVIOUS VALUE OF THE SUM OF THE RECIPROCALS       *
C     * (SUMREC).                                                     *
C     ****************************************************************
C
      SUMREC = SUMREC + (1.0 / X)
C
C     ****************************************************************
C     * ADD ONE TO THE COUNTER (K) TO INDICATE THAT ANOTHER DATA POINT *
C     * (X) HAS BEEN PROCESSED.                                       *
C     ****************************************************************
C
      K = K + 1
C
C     ****************************************************************
C     * IF ALL DATA POINTS (X) HAVE NOT BEEN PROCESSED, BRANCH BACK TO *
C     * GET ANOTHER; OTHERWISE, CONTINUE TO THE NEXT INSTRUCTION.     *
C     ****************************************************************
C
      IF (K .LT. N) GO TO 1
C
C     ****************************************************************
C     * STEP 4.  DIVIDE THE SUM (SUMREC) OF THE RECIPROCALS OF THE DATA *
C     * POINTS INTO THE NUMBER (N) OF DATA POINTS IN THE SERIES.  THIS *
C     * GIVES THE HARMONIC MEAN (HMEAN).                             *
C     ****************************************************************
C
      HMEAN = N / SUMREC
```

```
C
C
C    ************************************************************************
C    * PRINT OUT THE HARMONIC MEAN (HMEAN).                                *
C    ************************************************************************
C
     WRITE (6, 600) HMEAN
 600 FORMAT (1X,21HTHE HARMONIC MEAN IS , F10.3)
C
C    ************************************************************************
C    * STOP THE PROGRAM.                                                   *
C    ************************************************************************
C
     STOP
C
C    ************************************************************************
C    * INDICATE THE END OF THE SOURCE PROGRAM.                             *
C    ************************************************************************
C
     END
```

D. Example Problems

An automobile travels at the rate of 20 MPH from town A to town B, then returns from town B to town A at the rate of 30 MPH. The towns are ten miles apart. Using the previously developed computer program, calculate the average rate of speed in traveling from A to B and returning from B to A.

Example 1

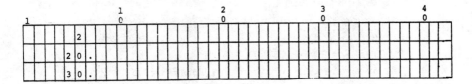

Input Data

THE HARMONIC MEAN IS 24.000

Solution

An airplane travels from Detroit, Michigan, to Miami, Florida, at an average speed of 620 miles per hours. Upon returning from Miami to Detroit, the pilot encounters a strong head wind and averages only 560 miles per hour. Find the average speed in the complete round trip.

Example 2

Input Data

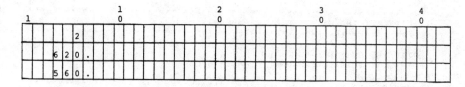

Solution   THE HARMONIC MEAN IS 588.475

☆ Calculation of a Geometric Mean

A. Statement of Problem

Given a set of N variates $X_1, X_2, \ldots, X_N$, using the formula

$$m_g = \sqrt[N]{X_1 \cdot X_2 \cdot \ldots \cdot X_N}$$

write a computer program to calculate the geometric mean of N variates.

B. Algorithm

1. Determine the number (K) of data points (P) for which the geometric mean is desired.
2. Insure that the number (K) of data points (P) is 1 or more.
3. Determine the product (PRODCT) of the K data points; that is, evaluate $P_1 \times P_2 \times \ldots \times P_K$. The value of any data point (P) must be greater than zero.
4. Calculate the geometric mean (GMEAN) by extracting the Kth root of the product (PRODCT) obtained in Step 3 above.

C. General Program

```
C
C      ***********************************************************************
C      * THIS PROGRAM COMPUTES THE GEOMETRIC MEAN OF A SERIES OF DATA       *
C      * POINTS.                                                            *
C      ***********************************************************************
C
C
C      ***********************************************************************
C      * STEP 1.  DETERMINE THE NUMBER (K) OF DATA POINTS (P) IN THE SER-*
C      * IES FOR WHICH THE GEOMETRIC MEAN IS DESIRED.                       *
C      ***********************************************************************
C
       READ (5, 500) K
  500  FORMAT (I6)
C
C      ***********************************************************************
C      * STEP 2.  TEST THE NUMBER (K) OF DATA POINTS (P) IN THE SERIES TO*
C      * INSURE IT IS 1 OR MORE; IF IT IS NOT, THE PROGRAM STOPS.           *
```

```
C     ******************************************************************
C
C        IF (K .LT. 1) STOP
C
C     ******************************************************************
C     * STEP 3.  EVALUATE THE PRODUCT OF THE DATA POINTS IN THE SERIES. *
C     ******************************************************************
C
C
C     ******************************************************************
C     * INITIALIZE THE PRODUCT (PRODCT) OF THE DATA POINTS BY SETTING  *
C     * IT EQUAL TO ONE.  THIS IS DONE SO THAT THE RESULT OF THE FIRST  *
C     * MULTIPLICATION OF PRODCT AND THE FIRST DATA POINT WILL EQUAL THE*
C     * VALUE OF THE FIRST DATA POINT.  NOTE THAT WE WILL BE CALCULATING*
C     * THE PRODUCT (PRODCT) IN A STEPWISE FASHION.  THAT IS, WE WILL   *
C     * CALCULATE THE PRODUCT BY MULTIPLYING THE VALUE OF THE PREVIOUS  *
C     * PRODUCT, INITIALIZED TO ONE, BY THE NEXT DATA POINT (P) AND RE- *
C     * PEATING THIS PROCESS K TIMES.                                  *
C     ******************************************************************
C
C        PRODCT = 1.0
C
C     ******************************************************************
C     * INITIALIZE THE COUNTER (J) BY SETTING IT EQUAL TO ZERO; THIS   *
C     * WILL BE USED TO COUNT THE NUMBER OF DATA POINTS (P) THAT HAVE   *
C     * BEEN PROCESSED.                                                *
C     ******************************************************************
C
C        J = 0
C
C     ******************************************************************
C     * GET THE NEXT DATA POINT (P).                                   *
C     ******************************************************************
C
   25    READ (5, 501) P
  501    FORMAT (F10.3)
C
C     ******************************************************************
C     * TEST THE DATA POINT (P) TO INSURE IT IS GREATER THAN ZERO; IF IT*
C     * IS NOT, THE PROGRAM STOPS.                                     *
C     ******************************************************************
C
C        IF (P .LE. 0.0) STOP
C
C     ******************************************************************
C     * DETERMINE THE PRODUCT OF THE CURRENT DATA POINT (P) AND THE PRE-*
C     * VIOUS VALUE OF THE PRODUCT (PRODCT) AND      STORE THE RESULT IN *
C     * PRODCT.                                                        *
C     ******************************************************************
C
C        PRODCT = PRODCT * P
C
C     ******************************************************************
C     * ADD ONE TO THE COUNTER (J) TO INDICATE THAT ANOTHER DATA POINT *
C     * (P) HAS BEEN PROCESSED.                                        *
C     ******************************************************************
C
C        J = J + 1
C
C     ******************************************************************
C     * IF ALL DATA POINTS (X) HAVE NOT BEEN PROCESSED, BRANCH BACK TO  *
C     * GET ANOTHER; OTHERWISE, CONTINUE TO THE NEXT INSTRUCTION.      *
C     ******************************************************************
C
C        IF (J .NE. K) GO TO 25
```

```
C
C
C      *****************************************************************
C      * STEP 4.  AT THIS POINT, THE PRODUCT OF ALL THE DATA POINTS HAS  *
C      * BEEN DETERMINED AND SAVED IN PRODCT.  EXTRACT THE KTH ROOT OF   *
C      * THE PRODUCT (PRODCT) GIVING THE GEOMETRIC MEAN (GMEAN).  NOTE   *
C      * THAT THE DOUBLE ASTERISKS (**) INDICATE  EXPONENTIATION AND THAT*
C      * RAISING A NUMBER TO THE POWER INDICATED BY THE RECIPROCAL OF K  *
C      * IS EQUIVALENT TO EXTRACTING THE KTH ROOT OF THAT NUMBER.        *
C      *****************************************************************
C
       GMEAN = PRODCT ** (1.0 / K)
C
C      *****************************************************************
C      * PRINT OUT THE GEOMETRIC MEAN OF THE SERIES (GMEAN).            *
C      *****************************************************************
C
       WRITE (6, 600) GMEAN
  600  FORMAT (1X,36HTHE GEOMETRIC MEAN OF THE SERIES IS , F10.3)
C
C      *****************************************************************
C      * STOP THE PROGRAM.                                             *
C      *****************************************************************
C
       STOP
C
C      *****************************************************************
C      * INDICATE THE END OF THE SOURCE PROGRAM.                       *
C      *****************************************************************
C
       END
```

D. Example Problems

Example 1   Given the variates 4, 11, 7, 8, and 15, use the computer program to calculate the geometric mean of these values.

Input Data

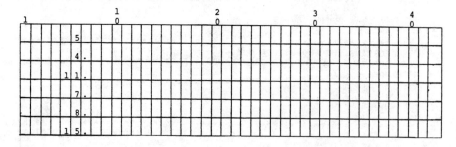

Solution   THE GEOMETRIC MEAN OF THE SERIES IS 8.195

Example 2   An investment firm deposits in its account the following dollar amounts in four years, as indicated:

Year	Amount of Deposit in Thousand Dollars
1979	18
1980	22
1981	25
1982	28

Find the average amount invested per year, using the geometric mean.

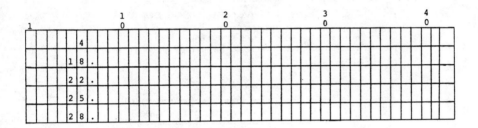

Input Data

THE GEOMETRIC MEAN OF THE SERIES IS 22.946

Solution

1. A computer salesman travels by automobile for a period of four days. If he travels 250 miles each day, but drives the first and last days at 50 miles per hour, the second at 52 miles per hour, and the third at 57 miles per hour, what is his average speed on the trip?

2. A school of business administration has 30 senior students attaining a grade point average of 2.85, a group of 45 junior students attaining a grade point average of 2.62, and a group of 58 sophomore students attaining a grade point average of 2.40. Compute the average grade point average for the three groups of students.

3. The number of felonies committed in a major U.S. city is indicated below:

Year	Number of Felonies
1979	1815
1980	2101
1981	2403
1982	2706

    Compute the geometric mean of the above data.

Chapter 3
Probability

A. Statement of Problem

Write a computer program which will evaluate $N!$, where N is a positive integer and is defined as:

$$N! = N \cdot (N - 1) \cdot \ldots \cdot 3 \cdot 2 \cdot 1$$

B. Algorithm

1. Determine the number (N) whose factorial (NFCTRL) is desired.

2. Test the number (N) whose factorial (NFCTRL) is desired to insure that it is greater than or equal to 0.

3. Calculate The factorial (NFCTRL) of the number (N).

   a. If the number (N) equals zero, the factorial (NFCTRL) equals one by definition.

   b. Evaluate ($1 \times 2 \times 3 \times \ldots \times N$). Multiply the number 1 by the first of the elements (NEXTNR) in the expression; then we

79

use the result of that operation as the multiplier of the second of the elements (NEXTNR) in the expression. Repeat the procedure of multiplying the intermediate product (IPRDCT) by the next number (NEXTNR) in the expression until the next number (NEXTNR) in the expression is greater than the number (N) whose factorial is desired.

c. The final value of the intermediate product (IPRDCT) is the value of the factorial (NFCTRL) of the number (N).

C. General Program

```
C
C
C    ******************************************************************
C    *  THIS PROGRAM CALCULATES THE FACTORIAL OF A NUMBER.           *
C    ******************************************************************
C
C
C
C    ******************************************************************
C    *  STEP 1.  DETERMINE THE NUMBER (N) WHOSE FACTORIAL (NFCTRL) IS *
C    *  DESIRED.                                                      *
C    ******************************************************************
C
     READ (5, 500) N
 500 FORMAT (I6)
C
C    ******************************************************************
C    *  STEP 2.  TEST THE NUMBER (N) WHOSE FACTORIAL (NFCTRL) IS DESIRED*
C    *  TO INSURE IT IS 0 OR GREATER.                                *
C    ******************************************************************
C
     IF (N .LT. 0) STOP
C
C    ******************************************************************
C    *  STEP 3.  CALCULATE THE FACTORIAL (NFCTRL) OF THE NUMBER (N).  *
C    ******************************************************************
C
C
C
C    ******************************************************************
C    *  STEP 3(A).  IF THE NUMBER (N) WHOSE FACTORIAL (NFCTRL) IS DE- *
C    *  SIRED EQUALS 1, THE FACTORIAL (NFCTRL) IS 1 BY DEFINITION.    *
C    ******************************************************************
C
C
C
C    ******************************************************************
C    *  IF THE VALUE OF THE NUMBER (N) WHOSE FACTORIAL (NFCTRL) IS DE- *
C    *  SIRED IS NOT EQUAL TO 0, BRANCH TO THE FIRST INSTRUCTION OF THE *
C    *  ROUTINE WHICH EVALUATES THE EXPRESSION (1 X 2 X 3 X . . . X N); *
C    *  OTHERWISE, CONTINUE TO THE NEXT INSTRUCTION.                  *
C    ******************************************************************
C
     IF (N .NE. 0) GO TO 10
C
C    ******************************************************************
C    *  AT THIS POINT IT HAS BEEN DETERMINED THAT THE NUMBER (N) WHOSE *
C    *  FACTORIAL (NFCTRL) IS DESIRED IS 0; SINCE, BY DEFINITION, THE *
C    *  FACTORIAL OF 0 IS 1, SET THE VALUE OF THE FACTORIAL (NFCTRL)  *
C    *  EQUAL TO 1.                                                   *
C    ******************************************************************
C
     NFCTRL = 1
```

```
C
C      *****************************************************************
C      * BRANCH TO THE INSTRUCTION THAT PRINTS OUT THE VALUE OF THE FAC- *
C      * TORIAL (NFCTRL).                                               *
C      *****************************************************************
C
       GO TO 40
C
C      *****************************************************************
C      * STEP 3(B).  EVALUATE THE EXPRESSION (1 X 2 X 3 X . . . X N).   *
C      *****************************************************************
C
C
C      *****************************************************************
C      * INITIALIZE THE INTERMEDIATE PRODUCT (IPRDCT) BY SETTING IT EQUAL *
C      * TO ONE;  THE PURPOSE OF THIS IS TO INSURE THAT THE FIRST MULTI- *
C      * PLICATION BELOW OF THE INTERMEDIATE PRODUCT (IPRDCT) TIMES THE  *
C      * FIRST NUMBER IN THE EXPRESSION (1 X 2 X 3 X . . . X N) WILL RE- *
C      * SULT IN AN ANSWER OF ONE.                                      *
C      *****************************************************************
   10  IPRDCT = 1
C
C      *****************************************************************
C      * INITIALIZE THE NEXT NUMBER (NEXTNR) BY SETTING IT EQUAL TO ZERO; *
C      * BY REPETITIVELY ADDING ONE TO THIS VARIABLE (NEXTNR) THE SUC-  *
C      * CESSIVE VALUES IN THE EXPRESSION (1 X 2 X 3 X . . . X N) CAN BE *
C      * DERIVED.                                                       *
C      *****************************************************************
C
       NEXTNR = 0
C
C      *****************************************************************
C      * ADD ONE TO THE NEXT NUMBER (NEXTNR) TO OBTAIN THE NEXT VALUE IN *
C      * THE EXPRESSION (1 X 2 X 3 X . . . X N).                        *
C      *****************************************************************
C
   20  NEXTNR = NEXTNR + 1
C
C      *****************************************************************
C      * IF THE CURRENT VALUE OF THE NEXT NUMBER (NEXTNR) IN THE EXPRES- *
C      * SION (1 X 2 X 3 X . . . X N) IS GREATER THAN THE NUMBER (N)    *
C      * WHOSE FACTORIAL (NFCTRL) IS DESIRED, THE EVALUATION OF THE FAC- *
C      * TORIAL (NFCTRL) HAS BEEN COMPLETED, AND A BRANCH OUT OF THE    *
C      * MULTIPLICATION PROCESS IS TAKEN.  OTHERWISE, THE PROGRAM CONTIN- *
C      * UES TO THE NEXT INSTRUCTION.                                   *
C      *****************************************************************
C
       IF (NEXTNR .GT. N) GO TO 30
C
C      *****************************************************************
C      * MULTIPLY THE CURRENT VALUE OF THE INTERMEDIATE PRODUCT (IPRDCT) *
C      * BY THE NEXT NUMBER (NEXTNR) IN THE EXPRESSION (1 X 2 X 3 X . . .*
C      * X N) GIVING A NEW VALUE TO THE INTERMEDIATE PRODUCT (IPRDCT).  *
C      *****************************************************************
C
       IPRDCT = IPRDCT * NEXTNR
C
C      *****************************************************************
C      * BRANCH BACK TO THE INSTRUCTION THAT GENERATES THE NEXT NUMBER  *
C      * (NEXTNR) IN THE EXPRESSION (1 X 2 X 3 X . . . X N).            *
C      *****************************************************************
C
       GO TO 20
```

```
C
C    ************************************************************************
C    * STEP 3(C).   AT THIS POINT THE CURRENT VALUE OF THE INTERMEDIATE  *
C    * PRODUCT (IPRDCT) CONTAINS THE VALUE OF THE FACTORIAL OF THE NUM-*
C    * BER (N); SET THE VALUE OF THE FACTORIAL (NFCTRL) EQUAL TO THE      *
C    * CURRENT VALUE OF THE INTERMEDIATE PRODUCT (IPRDCT).                *
C    ************************************************************************
C
  30    NFCTRL = IPRDCT
C
C    ************************************************************************
C    * PRINT OUT THE VALUE OF THE FACTORIAL (NFCTRL) OF THE NUMBER (N).*
C    ************************************************************************
C
  40    WRITE (6, 600) N, NFCTRL
 600    FORMAT (1X,13HTHE VALUE OF , I5, 14H FACTORIAL IS , I9)
C
C    ************************************************************************
C    * STOP THE PROGRAM.                                                 *
C    ************************************************************************
C
        STOP
C
C    ************************************************************************
C    * INDICATE THE END OF THE SOURCE PROGRAM.                           *
C    ************************************************************************
C
        END
```

D. Example Problems

Example 1   Evaluate 6!

Input Data
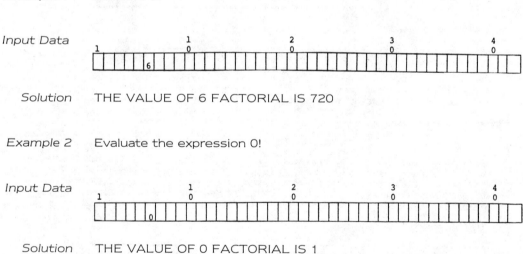

Solution   THE VALUE OF 6 FACTORIAL IS 720

Example 2   Evaluate the expression 0!

Input Data

Solution   THE VALUE OF 0 FACTORIAL IS 1

A. Statement of Problem

Write a computer program which determines the number of permutations P of n distinct objects taken r at a time as determined by the following formula:

$$_nP_r = \frac{n!}{(n-r)!}$$

B. Algorithm

1. Determine the values of n and r to be used in the evaluation of the given equation.

2. Insure that n and r are greater than zero.

3. Insure that the value of r is not greater than the value of n.

4. Determine the factorial of n (FCTRL1).

5. Determine the factorial of the quantity n minus r (FCTRL2).

6. Calculate the number of permutations (NRPERM) of the n things taken r at a time. This is accomplished by dividing the factorial of n (FCTRL1) by the factorial of the quantity n minus r (FCTRL2).

C. General Program

```
C
C
C    ************************************************************
C    * THIS PROGRAM CALCULATES THE NUMBER OF PERMUTATIONS OF N THINGS  *
C    * TAKEN R AT A TIME.                                       *
C    ************************************************************
C
C
C
C    ************************************************************
C    * INFORM THE COMPUTER THAT THE VARIABLES FCTRL1, FCTRL2, AND R    *
C    * WILL BE INTEGER VARIABLES RATHER THAN REAL VARIABLES.    *
C    ************************************************************
C
      INTEGER FCTRL1, FCTRL2, R
C
C
C    ************************************************************
C    * STEP 1.  DETERMINE THE VALUES OF N AND R TO BE USED IN THE EVAL-*
C    * UATION OF THE GIVEN EQUATION.                            *
C    ************************************************************
C
      READ (5, 500) N, R
  500 FORMAT (2I6)
C
C
C    ************************************************************
C    * STEP 2.  INSURE THAT N AND R ARE GREATER THAN ZERO.  IF EITHER N*
C    * OR R IS NOT GREATER THAN ZERO THE PROGRAM STOPS.         *
C    ************************************************************
C
      IF (N .LT. 1) STOP
      IF (R .LT. 1) STOP
```

83

```
C
C         *************************************************************
C         * STEP 3.  INSURE THAT THE VALUE OF R IS NOT GREATER THAN THE VAL-*
C         * UE OF N.  IF R IS GREATER THAN N THE PROGRAM STOPS.        *
C         *************************************************************
C
          IF (R .GT. N) STOP
C
C         *************************************************************
C         * STEP 4.  DETERMINE THE FACTORIAL OF N (FCTRL1) BY EVALUATING THE*
C         * EXPRESSION 1 X 2 X 3 X . . . X N.                          *
C         *************************************************************
C
C
C         *************************************************************
C         * INITIALIZE THE INTERMEDIATE PRODUCT (IPRDCT) BY SETTING IT EQUAL*
C         * TO ONE; THE PURPOSE OF THIS IS TO INSURE THAT THE FIRST MULTI- *
C         * PLICATION BELOW OF THE INTERMEDIATE PRODUCT TIMES THE FIRST NUM-*
C         * BER IN THE EXPRESSION 1 X 2 X 3 X . . . X N WILL RESULT IN AN  *
C         * ANSWER OF ONE.                                             *
C         *************************************************************
C
          IPRDCT = 1
C
C         *************************************************************
C         * INITIALIZE THE NEXT NUMBER (NEXTNR) BY SETTING IT EQUAL TO ZERO;*
C         * BY REPETITIVELY ADDING ONE TO THIS VARIABLE THE SUCCESSIVE VAL- *
C         * UES IN THE EXPRESSION 1 X 2 X 3 X . . . X N CAN BE DERIVED.    *
C         *************************************************************
C
          NEXTNR = 0
C
C         *************************************************************
C         * ADD ONE TO THE NEXT NUMBER (NEXTNR) TO OBTAIN THE NEXT VALUE IN *
C         * THE EXPRESSION 1 X 2 X 3 X . . . X N.                      *
C         *************************************************************
C
   20     NEXTNR = NEXTNR + 1
C
C         *************************************************************
C         * IF THE CURRENT VALUE OF THE NEXT NUMBER (NEXTNR) IN THE EXPRES- *
C         * SION 1 X 2 X 3 X . . . X N IS GREATER THAN THE NUMBER (N) WHOSE *
C         * FACTORIAL IS DESIRED, THE EVALUATION OF THE FACTORIAL HAS BEEN  *
C         * COMPLETED AND A BRANCH OUT OF THE MULTIPLICATION PROCESS IS TAK-*
C         * EN.  OTHERWISE, THE PROGRAM CONTINUES TO THE NEXT INSTRUCTION.  *
C         *************************************************************
C
          IF (NEXTNR .GT. N) GO TO 30
C
C         *************************************************************
C         * MULTIPLY THE CURRENT VALUE OF THE INTERMEDIATE PRODUCT (IPRDCT) *
C         * BY THE NEXT NUMBER (NEXTNR) IN THE EXPRESSION 1 X 2 X 3 X . . . *
C         * X N GIVING A NEW VALUE TO THE INTERMEDIATE PRODUCT.        *
C         *************************************************************
C
          IPRDCT = IPRDCT * NEXTNR
C
C         *************************************************************
C         * BRANCH BACK TO THE INSTRUCTION THAT GENERATES THE NEXT NUMBER  *
C         * (NEXTNR) IN THE EXPRESSION 1 X 2 X 3 X . . . X N.          *
C         *************************************************************
C
          GO TO 20
C
C         *************************************************************
C         * AT THIS POINT THE CURRENT VALUE OF THE INTERMEDIATE PRODUCT    *
```

```
C      * (IPRDCT) CONTAINS THE VALUE OF THE FACTORIAL OF N; SET THE FAC-  *
C      * TORIAL OF N (FCTRL1) EQUAL TO THE CURRENT VALUE OF THE INTER-     *
C      * MEDIATE PRODUCT (IPRDCT).                                         *
C      ****************************************************************
C
 30    FCTRL1 = IPRDCT
C
C      ****************************************************************
C      * STEP 5.  DETERMINE THE FACTORIAL OF THE QUANTITY N MINUS R       *
C      * (FCTRL2).                                                        *
C      ****************************************************************
C
C
C      ****************************************************************
C      * EVALUATE THE QUANTITY N MINUS R AND STORE THE RESULT IN THE VAR-*
C      * IABLE NLESSR.                                                    *
C      ****************************************************************
C
       NLESSR = N - R
C
C      ****************************************************************
C      * NOW DETERMINE THE FACTORIAL OF NLESSR AND STORE THE RESULT IN   *
C      * THE VARIABLE FCTRL2.  THE PROCEDURE TO DO THIS IS IDENTICAL TO  *
C      * THE ABOVE PROCEDURE TO DETERMINE THE FACTORIAL OF N.            *
C      ****************************************************************
C
       IPRDCT = 1
       NEXTNR = 0
 40    NEXTNR = NEXTNR + 1
       IF (NEXTNR .GT. NLESSR) GO TO 60
       IPRDCT = IPRDCT * NEXTNR
       GO TO 40
 60    FCTRL2 = IPRDCT
C
C      ****************************************************************
C      * STEP 6.  DETERMINE THE NUMBER OF PERMUTATIONS (NRPERM) BY DIVID-*
C      * ING THE FACTORIAL OF N (FCTRL1) BY THE FACTORIAL OF THE QUANTITY*
C      * N MINUS R (FCTRL2).                                             *
C      ****************************************************************
C
       NRPERM = FCTRL1 / FCTRL2
C
C      ****************************************************************
C      * PRINT OUT THE NUMBER OF PERMUTATIONS (NRPERM) OF N THINGS TAKEN *
C      * R AT A TIME.                                                    *
C      ****************************************************************
C
       WRITE (6, 600) N, R, NRPERM
 600   FORMAT (1X, I6, 8H THINGS , I6, 18H AT A TIME YIELDS , I6,
      *   13H PERMUTATIONS)
C
C      ****************************************************************
C      * STOP THE PROGRAM.                                               *
C      ****************************************************************
C
       STOP
C
C      ****************************************************************
C      * INDICATE THE END OF THE SOURCE PROGRAM.                         *
C      ****************************************************************
C
       END
```

D. Example Problems

Example 1   How many different arrangements of three books can be formed from a total of five textbooks?

Input Data

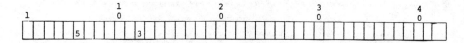

Solution   5 THINGS 3 AT A TIME YIELDS 60 PERMUTATIONS

Example 2   How many different five-letter actual and/or nonsense words can be made from the following set of letters?

A   G   B   D   O   Z   Y

Input Data

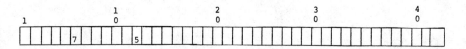

Solution   7 THINGS 5 AT A TIME YIELDS 2520 PERMUTATIONS

☆ Calculation of a Combination

A. Statement of Problem

Write a computer program which determines the number of combinations C of n distinct objects taken r at a time as determined by the following formula:

$$n^C r = \binom{n}{r} = \frac{n!}{r!\,(n - r)!}$$

B. Algorithm

1. Determine the values of n and r to be used in the evaluation of the given equation.
2. Insure that the values of n and r are greater than zero.
3. Insure that the value of r is not greater than the value of n.
4. Determine the factorial of n (FCTRL1).

5. Determine the factorial of the quantity n minus r (FCTRL2).

6. Determine the factorial of r (FCTRL3).

7. Calculate the number of combinations (NRCOMB) of n things taken r at a time by dividing the factorial of n (FCTRL1) by the product of the factorial of the quantity n minus r (FCTRL2) and the factorial of r (FCTRL3).

C. General Program

```
C
C      *********************************************************************
C      * THIS PROGRAM CALCULATES THE NUMBER OF COMBINATIONS OF N THINGS   *
C      * TAKEN R AT A TIME.                                               *
C      *********************************************************************
C
C
C      *********************************************************************
C      * INFORM THE COMPUTER THAT THE VARIABLES FCTRL1, FCTRL2, FCTRL3    *
C      * AND R WILL BE INTEGER VARIABLES RATHER THAN REAL VARIABLES.      *
C      *********************************************************************
C
       INTEGER FCTRL1, FCTRL2, FCTRL3, R
C
C      *********************************************************************
C      * STEP 1.  DETERMINE THE VALUES OF N AND R TO BE USED IN THE EVAL-*
C      * UATION OF THE GIVEN EQUATION.                                    *
C      *********************************************************************
C
       READ (5, 500) N, R
  500  FORMAT (2I6)
C
C      *********************************************************************
C      * STEP 2.  INSURE THAT N AND R ARE GREATER THAN ZERO.  IF EITHER N*
C      * OR R IS NOT GREATER THAN ZERO THE PROGRAM STOPS.                 *
C      *********************************************************************
C
       IF (N .LT. 1) STOP
       IF (R .LT. 1) STOP
C
C      *********************************************************************
C      * STEP 3.  INSURE THAT THE VALUE OF R IS NOT GREATER THAN THE VAL-*
C      * UE OF N.  IF R IS GREATER THAN N THE PROGRAM STOPS.              *
C      *********************************************************************
C
       IF (R .GT. N) STOP
C
C      *********************************************************************
C      * STEP 4.  DETERMINE THE FACTORIAL OF N (FCTRL1) BY EVALUATING THE*
C      * EXPRESSION 1 X 2 X 3 X . . . X N.                                *
C      *********************************************************************
C
C
C      *********************************************************************
C      * INITIALIZE THE INTERMEDIATE PRODUCT (IPRDCT) BY SETTING IT EQUAL*
C      * TO ONE; THE PURPOSE OF THIS IS TO INSURE THAT THE FIRST MULTI-  *
C      * PLICATION BELOW OF THE INTERMEDIATE PRODUCT TIMES THE FIRST NUM-*
C      * BER IN THE EXPRESSION 1 X 2 X 3 X . . . X N WILL RESULT IN AN    *
C      * ANSWER OF ONE.                                                   *
C      *********************************************************************
C
       IPRDCT = 1
```

```
C
C
C      ************************************************************************
C      * INITIALIZE THE NEXT NUMBER (NEXTNR) BY SETTING IT EQUAL TO ZERO;*
C      * BY REPETITIVELY ADDING ONE TO THIS VARIABLE THE SUCCESSIVE VAL- *
C      * UES IN THE EXPRESSION 1 X 2 X 3 X . . . X N CAN BE DERIVED.     *
C      ************************************************************************
C
       NEXTNR = 0
C
C      ************************************************************************
C      * ADD ONE TO THE NEXT NUMBER (NEXTNR) TO OBTAIN THE NEXT VALUE IN *
C      * THE EXPRESSION 1 X 2 X 3 X . . . X N.                          *
C      ************************************************************************
C
   20  NEXTNR = NEXTNR + 1
C
C      ************************************************************************
C      * IF THE CURRENT VALUE OF THE NEXT NUMBER (NEXTNR) IN THE EXPRES- *
C      * SION 1 X 2 X 3 X . . . X N IS GREATER THAN THE NUMBER (N) WHOSE *
C      * FACTORIAL IS DESIRED, THE EVALUATION OF THE FACTORIAL HAS BEEN  *
C      * COMPLETED AND A BRANCH OUT OF THE MULTIPLICATION PROCESS IS TAK-*
C      * EN.  OTHERWISE, THE PROGRAM CONTINUES TO THE NEXT INSTRUCTION.  *
C      ************************************************************************
C
       IF (NEXTNR .GT. N) GO TO 30
C
C      ************************************************************************
C      * MULTIPLY THE CURRENT VALUE OF THE INTERMEDIATE PRODUCT (IPRDCT) *
C      * BY THE NEXT NUMBER (NEXTNR) IN THE EXPRESSION 1 X 2 X 3 X . . . *
C      * X N GIVING A NEW VALUE TO THE INTERMEDIATE PRODUCT.            *
C      ************************************************************************
C
       IPRDCT = IPRDCT * NEXTNR
C
C      ************************************************************************
C      * BRANCH BACK TO THE INSTRUCTION THAT GENERATES THE NEXT NUMBER   *
C      * (NEXTNR) IN THE EXPRESSION 1 X 2 X 3 X . . . X N.              *
C      ************************************************************************
C
       GO TO 20
C
C      ************************************************************************
C      * AT THIS POINT THE CURRENT VALUE OF THE INTERMEDIATE PRODUCT     *
C      * (IPRDCT) CONTAINS THE VALUE OF THE FACTORIAL OF N; SET THE FAC- *
C      * TORIAL OF N (FCTRL1) EQUAL TO THE CURRENT VALUE OF THE INTER-   *
C      * MEDIATE PRODUCT (IPRDCT).                                      *
C      ************************************************************************
C
   30  FCTRL1 = IPRDCT
C
C      ************************************************************************
C      * STEP 5.  DETERMINE THE FACTORIAL OF THE QUANTITY N MINUS R      *
C      * (FCTRL2).                                                      *
C      ************************************************************************
C
C
C      ************************************************************************
C      * EVALUATE THE QUANTITY N MINUS R AND STORE THE RESULT IN THE VAR-*
C      * IABLE NLESSR.                                                  *
C      ************************************************************************
C
       NLESSR = N - R
C
C      ************************************************************************
C      * NOW DETERMINE THE FACTORIAL OF NLESSR AND STORE THE RESULT IN   *
```

```
C      * THE VARIABLE FCTRL2.  THE PROCEDURE TO DO THIS IS IDENTICAL TO  *
C      * THE ABOVE PROCEDURE TO DETERMINE THE FACTORIAL OF N.            *
C      ****************************************************************
C
       IPRDCT = 1
       NEXTNR = 0
 40    NEXTNR = NEXTNR + 1
       IF (NEXTNR .GT. NLESSR) GO TO 60
       IPRDCT = IPRDCT * NEXTNR
       GO TO 40
 60    FCTRL2 = IPRDCT
C
C      ****************************************************************
C      * STEP 6.  DETERMINE THE FACTORIAL OF R AND STORE THE RESULT IN  *
C      * THE VARIABLE FCTRL3.  THE PROCEDURE TO DO THIS IS IDENTICAL TO  *
C      * THE ABOVE PROCEDURE TO DETERMINE THE FACTORIAL OF N.            *
C      ****************************************************************
C
       IPRDCT = 1
       NEXTNR = 0
 80    NEXTNR = NEXTNR + 1
       IF (NEXTNR .GT. R) GO TO 100
       IPRDCT = IPRDCT * NEXTNR
       GO TO 80
 100   FCTRL3 = IPRDCT
C
C      ****************************************************************
C      * STEP 7.  DETERMINE THE NUMBER OF COMBINATIONS (NRCOMB) BY DIVID-*
C      * ING THE FACTORIAL OF N (FCTRL1) BY THE PRODUCT OF THE FACTORIAL *
C      * OF THE QUANTITY N MINUS (FCTRL2) AND THE FACTORIAL OF R         *
C      * (FCTRL3).                                                       *
C      ****************************************************************
C
       NRCOMB = FCTRL1 / (FCTRL2 * FCTRL3)
C
C      ****************************************************************
C      * PRINT OUT THE NUMBER OF COMBINATIONS (NRCOMB) OF N THINGS TAKEN *
C      * R AT A TIME.                                                    *
C      ****************************************************************
C
       WRITE (6, 600) N, R, NRCOMB
 600   FORMAT (1X, I6, 8H THINGS , I6, 18H AT A TIME YIELDS , I6,
      *   13H COMBINATIONS)
C
C      ****************************************************************
C      * STOP THE PROGRAM.                                              *
C      ****************************************************************
C
       STOP
C
C      ****************************************************************
C      * INDICATE THE END OF THE SOURCE PROGRAM.                        *
C      ****************************************************************
C
       END
```

D. Example Problems

Find the number of ways a committee of three can be selected from a group of 12 persons.

Example 1

Input Data

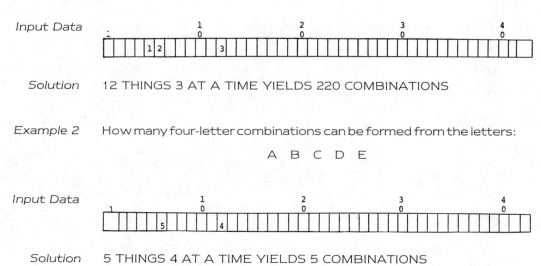

Solution   12 THINGS 3 AT A TIME YIELDS 220 COMBINATIONS

Example 2   How many four-letter combinations can be formed from the letters:

A B C D E

Input Data

Solution   5 THINGS 4 AT A TIME YIELDS 5 COMBINATIONS

☆ Expectation

A. Statement of Problem

Given a random variable, X, and its associated probability function as indicated below:

$f(x_i)$	$f(x_1)$	$f(x_2)$	...		$f(x_n)$
x_i	x_1	x_2	...		x_n

where $f(x_i)$ represents the probability of occurrence of each value of a random variable and n represents the number of pairs of outcomes; x_i represents the outcome or possible values of the random variable. Write a computer program which will compute the mathematical expectation of a random variable as given by the following formula:

$$E(X) = \sum_{i=1}^{n} x_i f(x_i)$$

B. Algorithm

1. Determine the number (N) of pairs of outcomes and associated probabilities (OUTCOM, PROB) to be used in the analysis.

2. Test the number (N) of pairs of outcomes and associated probabilities (OUTCOM, PROB) to insure that it is 1 or more.

3. Calculate the expected value of the event (EXP) by accumulating the expected value of each of the individual outcomes (XVAL) for all pairs of outcomes and associated probabilities (OUTCOM, PROB).

    a. Compute the expected value of each individual outcome (XVAL) by multiplying the outcome (OUTCOM) by its associated probability (PROB).

    b. Accumulate the expected value of each of the individual outcomes (XVAL).

C. General Program

```
C
C     ************************************************************
C     * THIS PROGRAM CALCULATES THE EXPECTED VALUE OF AN EVENT.       *
C     ************************************************************
C
C
C     ************************************************************
C     * STEP 1.  DETERMINE THE NUMBER (N) OF PAIRS OF OUTCOMES AND     *
C     * ASSOCIATED PROBABILITIES (OUTCOM, PROB) TO BE USED IN THE ANAL- *
C     * YSIS.                                                         *
C     ************************************************************
C
      READ (5, 500) N
  500 FORMAT (I6)
C
C     ************************************************************
C     * STEP 2.  TEST THE NUMBER (N) OF PAIRS OF OUTCOMES AND ASSOCIATED*
C     * PROBABILITIES (OUTCOM, PROB) TO INSURE THAT IT IS 1 OR MORE.  IF*
C     * IT IS NOT, THE PROGRAM STOPS.                                 *
C     ************************************************************
C
      IF (N .LT. 1) STOP
C
C     ************************************************************
C     * STEP 3.  CALCULATE THE EXPECTED VALUE OF THE EVENT (EXP) BY      *
C     * SUMMING THE EXPECTED VALUES OF EACH OF THE INDIVIDUAL OUTCOMES   *
C     * (XVAL) FOR ALL PAIRS OF OUTCOMES AND ASSOCIATED PROBABILITIES    *
C     * (OUTCOM, PROB).                                               *
C     ************************************************************
C
C
C     ************************************************************
C     * INITIALIZE THE ACCUMULATOR (EXP) USED TO SUM THE EXPECTED VALUES*
C     * OF EACH OUTCOME (XVAL) BY SETTING IT EQUAL TO ZERO.           *
C     ************************************************************
C
      EXP = 0.0
C
C     ************************************************************
C     * INITIALIZE THE COUNTER (J) BY SETTING IT EQUAL TO ZERO.  THIS   *
C     * WILL BE USED TO COUNT THE NUMBER OF PAIRS OF OUTCOMES AND ASSOC-*
C     * IATED PROBABILITIES (OUTCOM, PROB) THAT HAVE BEEN PROCESSED.    *
```

```
C     ******************************************************************
C
C           J = 0
C
C     ******************************************************************
C     * GET THE NEXT PAIR OF OUTCOMES AND ASSOCIATED PROBABILITIES     *
C     * (OUTCOM, PROB).                                                *
C     ******************************************************************
C
  15      READ (5, 501) OUTCOM, PROB
 501      FORMAT (F10.3, F10.4)
C
C     ******************************************************************
C     * STEP 3(A).  CALCULATE THE EXPECTED VALUE (XVAL) OF THIS OUTCOME *
C     * (OUTCOM) BY MULTIPLYING THE OUTCOME (OUTCOM) BY ITS ASSOCIATED  *
C     * PROBABILITY (PROB).                                            *
C     ******************************************************************
C
          XVAL = OUTCOM * PROB
C
C     ******************************************************************
C     * STEP 3(B).  ACCUMULATE THE EXPECTED VALUES (XVAL) OF THE OUT-  *
C     * COME.                                                          *
C     ******************************************************************
C
          EXP = EXP + XVAL
C
C     ******************************************************************
C     * ADD ONE TO THE COUNTER (J) TO INDICATE THAT ANOTHER PAIR OF OUT-*
C     * COMES AND ASSOCIATED PROBABILITIES (OUTCOM, PROB) HAS BEEN PRO- *
C     * CESSED.                                                        *
C     ******************************************************************
C
          J = J + 1
C
C     ******************************************************************
C     * IF ALL PAIRS OF OUTCOMES AND ASSOCIATED PROBABILITIES (OUTCOM, *
C     * PROB) HAVE NOT BEEN PROCESSED, BRANCH BACK TO GET ANOTHER PAIR; *
C     * OTHERWISE, CONTINUE TO THE NEXT INSTRUCTION.                   *
C     ******************************************************************
C
          IF (J .NE. N) GO TO 15
C
C     ******************************************************************
C     * PRINT OUT THE EXPECTED VALUE OF THE EVENT (EXP).               *
C     ******************************************************************
C
          WRITE (6, 600) EXP
 600      FORMAT (1X,35HTHE EXPECTED VALUE OF THE EVENT IS , F10.3)
C
C     ******************************************************************
C     * STOP THE PROGRAM.                                             *
C     ******************************************************************
C
          STOP
C
C     ******************************************************************
C     * INDICATE THE END OF THE SOURCE PROGRAM.                       *
C     ******************************************************************
C
          END
```

D. Example Problems

An ordinary six-faced die is thrown once. The probability function   *Example 1*
associated with the number of dots appearing on the top face of this
die is as follows:

$f(x_i)$	⅙	⅙	⅙	⅙	⅙	⅙
x_i	1	2	3	4	5	6

Compute the mathematical expectation of this random variable.

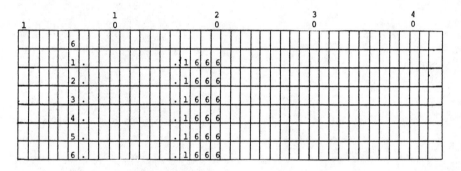   *Input Data*

THE EXPECTED VALUE OF THE EVENT IS 3.499   *Solution*

The number of automotive accidents in Painesville, Ohio, on any given   *Example 2*
day is 0, 1, 2, 3, 4, and 5 with corresponding probabilities of 0.37,
0.34, 0.18, 0.08, 0.02, and 0.01. What is the expected number of
accidents on any given day?

Input Data

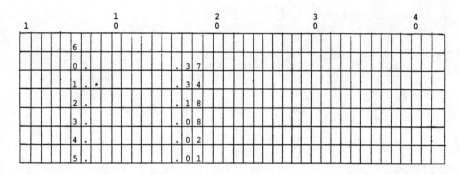

Solution   THE EXPECTED VALUE OF THE EVENT IS 1.070

1. Evaluate 7!

2. Find how many different three-letter words can be made from the letters of the word *racquet*.

3. Determine the number of ways in which three novels can be read from a total of six novels.

4. Assume that the expected value of any function, *H*, can be evaluated as

$$E(H) = \sum_{i=1}^{t} f(x_i)H(x_i)$$

Write a computer program which will evaluate:

    a. $E(X^2)$

    b. $E(2X - 1)$

    c. $E(3X^2 - 8)$

    for the following probability function:

$f(x_i)$	.2	.4	.3	.1
$X = x_i$	0	1	2	3

Chapter 4
Probability Distributions

☆ Binomial Distribution

A. Statement of Problem

Given a binomial experiment that can result in a success with a probability p and a failure with probability $q = 1 - p$. The probability distribution of the binomial random variable X (the number of successes in n independent trials) is defined as

$$P(X = x) = b(x, n, p)$$
$$= \binom{n}{x} p^x q^{n-x}, \qquad x = 0, 1, 2, \ldots, n$$

Write a computer program that computes (a) the probability of obtaining exactly x successes in n trials for a binomial experiment; and (b) the mean and variance of the binomial distribution as given by $m = np$ and $\sigma^2 = npq$, respectively.

B. Algorithm

1. Determine the number of independent trials (N) and the probability of a success (P).

2. Insure that the number of independent trials (N) is greater than zero and that the probability of a success (P) is within the range zero to one.

3. Compute the mean (BMEAN) and the variance (BVAR) of the distribution as indicated in the given formulas.

4. Determine the probability distribution of the binomial random variable (X). This is accomplished by repeatedly executing Steps 4a and 4b with values of X of 0, 1, ..., N.

   a. Determine the number of combinations $\binom{n}{x}$ as required by the formula.

   b. Evaluate the formula $\binom{n}{x} p^x q^{n-x}$ giving the probability (B) of X successes in N trials.

C. General Program

```
C
C     ****************************************************************
C     * THIS PROGRAM COMPUTES THE MEAN AND VARIANCE OF A BINOMIAL     *
C     * DISTRIBUTION AS WELL AS THE PROBABILITY DISTRIBUTION OF THE BI- *
C     * NOMIAL RANDOM VARIABLE.                                        *
C     ****************************************************************
C
C
C
C     ****************************************************************
C     * INFORM THE COMPUTER THAT X, FCTRL1, FCTRL2 AND FCTRL3 WILL BE  *
C     * INTEGER VARIABLES RATHER THAN REAL VARIABLES.                  *
C     ****************************************************************
C
      INTEGER X, FCTRL1, FCTRL2, FCTRL3
C
C     ****************************************************************
C     * STEP 1.  DETERMINE THE NUMBER OF INDEPENDENT TRIALS (N) AND THE *
C     * PROBABILITY OF A SUCCESS (P).                                  *
C     ****************************************************************
C
      READ (5, 500) N, P
  500 FORMAT (I6, F10.3)
C
C     ****************************************************************
C     * INSURE THAT THE VALUES OF N AND P ARE WITHIN THE ALLOWABLE RANG-*
C     * ES.                                                            *
C     ****************************************************************
C
      IF (N .LE. 0) STOP
      IF (P .LT. 0.0 .OR. P .GT. 1.0) STOP
C
C     ****************************************************************
C     * STEP 3.  COMPUTE THE MEAN (BMEAN) AND VARIANCE (BVAR) OF THE    *
C     * DISTRIBUTION.                                                  *
```

```
C      *****************************************************************
C
       BMEAN = N * P
       WRITE (6, 600) BMEAN
  600  FORMAT (1X,41HTHE MEAN CF THE BINOMIAL DISTRIBUTION IS , F10.3)
       Q = 1.0 - P
       BVAR = N * P * Q
       WRITE (6, 601) BVAR
  601  FORMAT (1X,45HTHE VARIANCE OF THE BINOMIAL DISTRIBUTION IS ,
      *   F10.3)
C
C
C      *****************************************************************
C      * STEP 4.  CACLULATE THE PROBABILITY DISTRIBUTION OF THE BINOMIAL *
C      * RANDOM VARIABLE (X) FOR X = 0, 1, . . ., N.                   *
C      *****************************************************************
C
C
C
C      *****************************************************************
C      * SET X TO AN INITIAL VALUE OF ZERO.                            *
C      *****************************************************************
C
       X = 0
C
C
C      *****************************************************************
C      * STEP 4(A).  DETERMINE THE NUMBER OF COMBINATIONS OF N THINGS  *
C      * TAKEN X AT A TIME.  THE PROCEDURE TO DO THIS IS VERY SIMILAR TO *
C      * THE EARLIER PROGRAM WHICH CALCULATED THE NUMBER OF COMBINATIONS.*
C      *****************************************************************
C
  5    IPRDCT = 1
       NEXTNR = 0
  20   NEXTNR = NEXTNR + 1
       IF (NEXTNR .GT. N) GO TO 30
       IPRDCT = IPRDCT * NEXTNR
       GO TO 20
  30   FCTRL1 = IPRDCT
       NLESSX = N - X
       IPRDCT = 1
       NEXTNR = 0
  40   NEXTNR = NEXTNR + 1
       IF (NEXTNR .GT. NLESSX) GO TO 60
       IPRDCT = IPRDCT * NEXTNR
       GO TO 40
  60   FCTRL2 = IPRDCT
       IPRDCT = 1
       NEXTNR = 0
  80   NEXTNR = NEXTNR + 1
       IF (NEXTNR .GT. X) GO TO 100
       IPRDCT = IPRDCT * NEXTNR
       GO TO 80
  100  FCTRL3 = IPRDCT
C
C      *****************************************************************
C      * STEP 4(B).  EVALUATE THE GIVEN FORMULA.  CALCULATE THE PROBA- *
C      * BILITY (B) OF X SUCCESSES IN N TRIALS.                        *
C      *****************************************************************
C
       B = (FCTRL1 / (FCTRL2 * FCTRL3)) * (P ** X) * (Q ** (N - X))
       WRITE (6, 602) X, N, B
  602  FORMAT (1X,19HTHE PROBABILITY OF , I6, 14H SUCCESSES IN , I6,
      *   11H TRIALS IS , F10.3)
C
C      *****************************************************************
C      * ADD 1 TO THE NUMBER OF SUCCESSES (X).                         *
```

```
C     *********************************************************************
C
      X = X + 1
C
C     *********************************************************************
C     * IF THE PROBABILITY FOR ALL SUCCESSES (X) HAS NOT BEEN EVALUATED *
C     * BRANCH BACK TO THE ROUTINE TO EVALUATE THE PROBABILITY FOR THE  *
C     * CURRENT NUMBER OF SUCCESSES (X).                                *
C     *********************************************************************
C
      IF (X .LE. N) GO TO 5
      STOP
      END
```

D. General Problems

A leading goalie in the NHL allows 5 percent of the opponents' shots in the nets for a goal. If in a particular game an opposing team takes 10 shots on goal, find the probability that the goalie allows exactly two goals during the game. (Assume the shots are all independent.)

Example 1

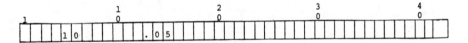

Input Data

THE MEAN OF THE BINOMIAL DISTRIBUTION IS 0.500
THE VARIANCE OF THE BINOMIAL DISTRIBUTION IS 0.475
THE PROBABILITY OF 0 SUCCESSES IN 10 TRIALS IS 0.599
THE PROBABILITY OF 1 SUCCESSES IN 10 TRIALS IS 0.315
THE PROBABILITY OF 2 SUCCESSES IN 10 TRIALS IS 0.075
THE PROBABILITY OF 3 SUCCESSES IN 10 TRIALS IS 0.010
THE PROBABILITY OF 4 SUCCESSES IN 10 TRIALS IS 0.001
THE PROBABILITY OF 5 SUCCESSES IN 10 TRIALS IS 0.000
THE PROBABILITY OF 6 SUCCESSES IN 10 TRIALS IS 0.000
THE PROBABILITY OF 7 SUCCESSES IN 10 TRIALS IS 0.000
THE PROBABILITY OF 8 SUCCESSES IN 10 TRIALS IS 0.000
THE PROBABILITY OF 9 SUCCESSES IN 10 TRIALS IS 0.000
THE PROBABILITY OF 10 SUCCESSES IN 10 TRIALS IS 0.000

Solution

The probability that a person recovers from a rattlesnake bite is 0.2. If five individuals have been inflicted with the bite, find the probability that exactly three individuals will recover from the bites.

Example 2

Input Data

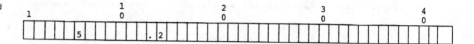

Solution THE MEAN OF THE BINOMIAL DISTRIBUTION IS 1.000
THE VARIANCE OF THE BINOMIAL DISTRIBUTION IS 0.800
THE PROBABILITY OF 0 SUCCESSES IN 5 TRIALS IS 0.328
THE PROBABILITY OF 1 SUCCESSES IN 5 TRIALS IS 0.410
THE PROBABILITY OF 2 SUCCESSES IN 5 TRIALS IS 0.205
THE PROBABILITY OF 3 SUCCESSES IN 5 TRIALS IS 0.051
THE PROBABILITY OF 4 SUCCESSES IN 5 TRIALS IS 0.006
THE PROBABILITY OF 5 SUCCESSES IN 5 TRIALS IS 0.000

☆ Cumulative Binomial

A. Statement of Problem

The probability of obtaining x_i or fewer successes in n trials is given by the equation

$$P(X \leqslant x_i) = b(x; n, p)$$

$$= \sum_{x=0}^{x_i} \binom{n}{x} p^x q^{n-x}$$

Write a computer program which will compute (a) the probability that the binomial random variable X falls at or below a certain value of x_i and (b) the mean and the variance of the binomial distribution.

B. Algorithm

1. Determine the number of independent trials (N) and the probability of a success (P).

2. Insure that the number of independent trials (N) is greater than zero and that the probability of a success (P) is within the range zero to one.

3. Compute the mean (BMEAN) and the variance (BVAR) of the distribution as indicated in the given formulas.

4. Determine the cumulative probability of the binomial random variable (X) by repeatedly executing Steps 4a through 4c with values of X of 0, 1,..., N.

a. Determine the number of combinations $\binom{n}{x}$ as required by the formula.

b. Evaluate the formula $\binom{n}{x}p^{x}q^{n-x}$, giving the probability (B) of X successes in N trials.

c. Accumulate the sum of the probabilities derived in Step 4b. Print this sum on each iteration of Step 4c.

C. General Program

```
C
C
C     ****************************************************************
C     * THIS PROGRAM COMPUTES THE MEAN AND VARIANCE OF A BINOMIAL    *
C     * DISTRIBUTION AS WELL AS THE CUMULATIVE PROBABILITY DISTRIBUTION *
C     * OF THE BINOMIAL RANDOM VARIABLE.                             *
C     ****************************************************************
C
C
C
C     ****************************************************************
C     * STEPS 1 THROUGH 3 OF THIS PROGRAM ARE IDENTICAL TO THOSE OF THE *
C     * PRECEDING  PROGRAM AND THUS WILL BE GIVEN WITHOUT COMMENT.   *
C     ****************************************************************
C
      INTEGER X, FCTRL1, FCTRL2, FCTRL3
      READ (5, 500) N, P
  500 FORMAT (I6, F10.3)
      IF (N .LE. 0) STOP
      IF (P .LT. 0.0 .OR. P .GT. 1.0) STOP
      BMEAN = N * P
      WRITE (6, 600) BMEAN
  600 FORMAT (1X,41HTHE MEAN OF THE BINOMIAL DISTRIBUTION IS , F10.3)
      Q = 1.0 - P
      BVAR = N * P * Q
      WRITE (6, 601) BVAR
  601 FORMAT (1X,45HTHE VARIANCE OF THE BINOMIAL DISTRIBUTION IS ,
     *  F10.3)
C
C
C     ****************************************************************
C     * STEP 4.   CALCULATE THE CUMULATIVE PROBABILITY DISTRIBUTION OF *
C     * BINOMIAL RANDOM VARIABLE (X) FOR X = 0, 1, . . ., N.         *
C     ****************************************************************
C
C
C
C     ****************************************************************
C     * SET THE BINOMIAL RANDOM VARIABLE (X) TO AN INITIAL VALUE OF 0. *
C     ****************************************************************
C
      X = 0
C
C     ****************************************************************
C     * INITIALIZE CUMB TO ZERO.  THIS VARIABLE WILL BE USED TO ACCUMU- *
C     * LATE THE SUM OF THE PROBABILITIES OF X SUCCESSES IN N TRIALS FOR *
C     * X = 0, 1, . . ., N.                                          *
C     ****************************************************************
C
      CUMB = 0.0
C
C     ****************************************************************
C     * STEP 4(A).   DETERMINE THE NUMBER OF COMBINATIONS OF N THINGS *
C     * TAKEN X AT A TIME AS REQUIRED BY THE GIVEN FORMULA.          *
C     ****************************************************************
C
```

```
C
C       ********************************************************************
C       * INITIALIZE THE INTERMEDIATE PRODUCT (IPRDCT) BY SETTING IT EQUAL*
C       * TO ONE; THE PURPOSE OF THIS IS TO INSURE THAT THE FIRST MULTI-  *
C       * PLICATION BELOW OF THE INTERMEDIATE PRODUCT TIMES THE FIRST NUM-*
C       * BER IN THE EXPRESSION 1 X 2 X 3 X . . . N WILL RESULT IN AN     *
C       * ANSWER OF ONE.                                                  *
C       ********************************************************************
C
  5     IPRDCT = 1
C
C       ********************************************************************
C       * REPETITIVELY CYCLE THROUGH THE SET OF INSTRUCTIONS FROM THE ONE *
C       * FOLLOWING THE DO STATEMENT THROUGH AND INCLUDING THE STATEMENT  *
C       * NUMBERED 20.  THE INDEX OF THE DO STATEMENT, THE VARIABLE NAMED *
C       * NEXTNR, WILL HAVE AN INITIAL VALUE OF ONE, AND WILL BE INCRE-   *
C       * MENTED BY ONE AFTER EACH CYCLE THROUGH THE SET OF INSTRUCTIONS  *
C       * MENTIONED.  THUS THE VALUE OF NEXTNR WILL BE ONE ON THE FIRST   *
C       * CYCLE, TWO ON THE SECOND CYCLE, . . ., AND N ON THE N-TH CYCLE. *
C       ********************************************************************
C
        DO 20 NEXTNR = 1, N
C
C       ********************************************************************
C       * MULTIPLY THE VALUE OF IPRDCT BY THE VALUE OF NEXTNR AND STORE   *
C       * THE RESULT IN IPRDCT.                                          *
C       ********************************************************************
C
        IPRDCT = IPRDCT * NEXTNR
C
C       ********************************************************************
C       * THE NEXT STATEMENT IS THE OBJECT OF THE EARLIER DO STATEMENT AND*
C       * IS THE LAST STATEMENT IN THE SET OF INSTRUCTIONS TO BE CYCLED   *
C       * THROUGH.  THIS STATEMENT PERFORMS NO CALCULATIONS.             *
C       ********************************************************************
C
  20    CONTINUE
C
C       ********************************************************************
C       * WHEN THE ABOVE SET OF INSTRUCTIONS HAS BEEN CYCLED THROUGH N    *
C       * TIMES THE PROGRAM CONTINUES WITH THE FOLLOWING STATEMENT.  THIS *
C       * STATEMENT STORES THE VALUE OF IPRDCT, WHICH EQUALS THE FACTORI- *
C       * AL OF N, IN THE VARIABLE FCTRL1.                               *
C       ********************************************************************
C
        FCTRL1 = IPRDCT
C
C       ********************************************************************
C       * THE ABOVE PROCEDURE WILL NOW BE REPEATED TO CALCULATE THE FACT- *
C       * ORIAL OF N - X (NLESSX) AND X.  ONLY ONE MODIFICATION TO THE    *
C       * PROCEDURE IS NECESSARY.  THIS INVOLVES TESTING THE VALUE OF THE *
C       * VARIABLE WHOSE FACTORIAL IS BEING CALCULATED TO INSURE IT IS    *
C       * GREATER THAN ZERO.                                             *
C       ********************************************************************
C
C
C
C       ********************************************************************
C       * CALCULATE N MINUS X.                                           *
C       ********************************************************************
C
        NLESSX = N - X
C
C       ********************************************************************
C       * CALCULATE THE FACTORIAL OF NLESSX.                             *
```

```
C      ************************************************************
C
       IPRDCT = 1
C
C      ************************************************************
C      * INSURE THAT NLESSX IS GREATER THAN ZERO.  IF IT IS NOT, BRANCH *
C      * TO THE STATEMENT NUMBERED 35.  STATEMENT 35 ASSUMES THAT THE   *
C      * FACTORIAL OF NLESSX IS STORED IN IPRDCT.  NOTICE THAT IF NLESSX *
C      * IS ZERO, ITS FACTORIAL WILL INDEED BY STORED IN IPRDCT SINCE THE*
C      * FACTORIAL OF ZERO IS ONE BY DEFINITION AND IPRDCT WAS INITIAL-  *
C      * IZED TO ONE.                                                    *
C      ************************************************************
C
       IF (NLESSX .EQ. 0) GO TO 35
C
C      ************************************************************
C      * NOTE THE USE OF NLESSX IN THE NEXT DO STATEMENT.  SINCE THIS    *
C      * VARIABLE CONTROLS THE NUMBER OF TIMES THE SET OF INSTRUCTIONS   *
C      * WILL BE CYCLED THROUGH, THE SET WILL BE EXECUTED NLESSX TIMES.  *
C      ************************************************************
C
       DO 30 NEXTNR = 1, NLESSX
       IPRDCT = IPRDCT * NEXTNR
   30  CONTINUE
C
C      ************************************************************
C      * STORE THE FACTORIAL OF NLESSX IN FCTRL2.                       *
C      ************************************************************
C
   35  FCTRL2 = IPRDCT
C
C      ************************************************************
C      * REPEAT THE ABOVE PROCEDURE TO CALCULATE THE FACTORIAL OF X; THE *
C      * FACTORIAL SHOULD BE STORED IN FCTRL3.                          *
C      ************************************************************
C
       IPRDCT = 1
       IF (X .EQ. 0) GO TO 45
       DO 40 NEXTNR = 1, X
       IPRDCT = IPRDCT * NEXTNR
   40  CONTINUE
   45  FCTRL3 = IPRDCT
C
C      ************************************************************
C      * STEP 4(B).  EVALUATE THE GIVEN FORMULA TO CALCULATE THE PROBA- *
C      * BILITY OF X SUCCESSES IN N TRIALS.                            *
C      ************************************************************
C
       B = (FCTRL1 / (FCTRL2 * FCTRL3)) * (P ** X) * (Q ** (N - X))
C
C      ************************************************************
C      * STEP 4(C).  ACCUMULATE THE PROBABILITIES (B) GIVING THE CUMULA- *
C      * TIVE PROBABILITY (CUMB).                                       *
C      ************************************************************
C
       CUMB = CUMB + B
       WRITE (6, 602) X, N, CUMB
  602  FORMAT (1X,19HTHE PROBABILITY OF , I6, 22H SUCCESSES OR LESS IN ,
      *  I6, 11H TRIALS IS , F10.3)
C
C      ************************************************************
C      * ADD 1 TO THE NUMBER OF SUCCESSES.                             *
C      ************************************************************
C
       X = X + 1
```

```
C
C
C      ******************************************************************
C      * IF THE PROBABILITY FOR ALL SUCCESSES (X) HAS NOT BEEN CALCULATED*
C      * BRANCH BACK TO THE ROUTINE WHICH CALCULATES THE REQUIRED PROBA- *
C      * BILITY.                                                         *
C      ******************************************************************
C
       IF (X .LE. N) GO TO 5
       STOP
       END
```

D. Example Problems

Example 1   Compute the cumulative binomial distribution for the binomial random variable X where $n = 5$ and $p = 0.6$.

Input Data

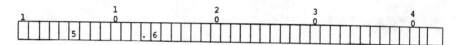

Solution   THE MEAN OF THE BINOMIAL DISTRIBUTION IS 3.000

THE VARIANCE OF THE BINOMIAL DISTRIBUTION IS 1.200

THE PROBABILITY OF 0 SUCCESSES OR LESS IN 5 TRIALS IS 0.010

THE PROBABILITY OF 1 SUCCESSES OR LESS IN 5 TRIALS IS 0.087

THE PROBABILITY OF 2 SUCCESSES OR LESS IN 5 TRIALS IS 0.317

THE PROBABILITY OF 3 SUCCESSES OR LESS IN 5 TRIALS IS 0.663

THE PROBABILITY OF 4 SUCCESSES OR LESS IN 5 TRIALS IS 0.922

THE PROBABILITY OF 5 SUCCESSES OR LESS IN 5 TRIALS IS 1.000

Example 2   Find the cumulative binomial distribution for the binomial random variable X with parameters $n = 0$ and $p = 0.4$.

Input Data

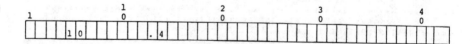

Solution   THE MEAN OF THE BINOMIAL DISTRIBUTION IS 4.000

THE VARIANCE OF THE BINOMIAL DISTRIBUTION IS 2.400

THE PROBABILITY OF 0 SUCCESSES OR LESS IN 10 TRIALS IS 0.006

THE PROBABILITY OF 1 SUCCESSES OR LESS IN 10 TRIALS IS 0.046

THE PROBABILITY OF 2 SUCCESSES OR LESS IN 10 TRIALS IS 0.167

THE PROBABILITY OF 3 SUCCESSES OR LESS IN 10 TRIALS IS 0.382
THE PROBABILITY OF 4 SUCCESSES OR LESS IN 10 TRIALS IS 0.633
THE PROBABILITY OF 5 SUCCESSES OR LESS IN 10 TRIALS IS 0.834
THE PROBABILITY OF 6 SUCCESSES OR LESS IN 10 TRIALS IS 0.945
THE PROBABILITY OF 7 SUCCESSES OR LESS IN 10 TRIALS IS 0.988
THE PROBABILITY OF 8 SUCCESSES OR LESS IN 10 TRIALS IS 0.998
THE PROBABILITY OF 9 SUCCESSES OR LESS IN 10 TRIALS IS 1.000
THE PROBABILITY OF 10 SUCCESSES OR LESS IN 10 TRIALS IS 1.000

☆ Multinomial Distribution

A. Statement of Problem

Suppose that a given experiment results in k outcomes E_1, E_2,...,
E_k, with corresponding probabilities P_1, P_2,..., P_k. Then the probability
distribution of the random variables x_1, x_2,..., x_k indicating the frequency of occurrences for E_1, E_2,..., E_k in n independent trials is
given by:

$$F(x_1, x_2, \ldots, x_k; p_1, p_2, \ldots, p_k, n) = \frac{n!}{x_1! x_2! \ldots x_k!} p_1^{x_1} p_2^{x_2} \ldots p_k^{x_k}$$

$$\text{where} \quad \sum_{i=1}^{k} x_i = n \quad \text{and} \quad \sum_{i=1}^{k} p_i = 1$$

Write a program which would evaluate the multinomial distribution
for k possible outcomes of events $E_1, E_2, \ldots, E_k$.

B. Algorithm

1. Determine the number of independent trials (N) and the
   number of possible outcomes (K). The number of independent trials must be equal to or greater than 1, and the number
   of possible outcomes must be 3 or greater.
2. Calculate the factorial of the number of independent trials (N).
3. Read in the K sets of numbers of occurrences of outcomes
   and their associated probabilities (X, P). As these are read,
   calculate the factorial of X, and P raised to the X power;
   derive the required products.
4. Calculate the multinomial probability (PROB) by dividing the
   factorial of N (NFACT) by the product of the factorials of X

(PRODX). Multiply this result by the product of the probabilities raised to the corresponding X values (PRODPX).

C. General Program

```
C
C
C      ******************************************************************
C      * THIS PROGRAM COMPUTES THE PROBABILITY OF A SET OF OUTCOMES IN A *
C      * MULTINOMIAL EXPERIMENT.                                         *
C      ******************************************************************
C
       INTEGER X, PRODX
C
C      ******************************************************************
C      * STEP 1.  DETERMINE THE NUMBER OF INDEPENDENT TRIALS (N) AND THE *
C      * NUMBER OF POSSIBLE OUTCOMES (K).  INSURE THAT EACH IS WITHIN ITS*
C      * PROPER LIMITS.                                                  *
C      ******************************************************************
C
       READ (5, 500) N, K
  500  FORMAT (2I6)
       IF (N .LT. 1) STOP
       IF (K .LT. 3) STOP
C
C      ******************************************************************
C      * GENERATE THE FIRST LINE OF OUTPUT.                             *
C      ******************************************************************
C
       WRITE (6, 600) N, K
  600  FORMAT (1X,35HIN THE MULTINOMIAL EXPERIMENT WITH , I6,
      *    12H TRIALS AND , I6, 30H OUTCOMES, THE PROBABILITY OF )
C
C      ******************************************************************
C      * STEP 2.  CALCULATE THE FACTORIAL OF N.                         *
C      ******************************************************************
C
       IPRDCT = 1
C
C      ******************************************************************
C      * REPETITIVELY CYCLE THROUGH THE SET OF INSTRUCTIONS FROM THE ONE *
C      * FOLLOWING THE DO STATEMENT THROUGH AND INCLUDING THE STATEMENT  *
C      * NUMBERED 10.  THE INDEX OF THE DO STATEMENT, THE VARIABLE NAMED *
C      * NEXTNR, WILL HAVE AN INITIAL VALUE OF ONE, AND WILL BE INCRE-   *
C      * MENTED BY ONE AFTER EACH CYCLE THROUGH THE SET OF INSTRUCTIONS  *
C      * MENTIONED.  THUS THE VALUE OF NEXTNR WILL BE ONE ON THE FIRST   *
C      * CYCLE, TWO ON THE SECOND CYCLE, . . ., AND N ON THE N-TH CYCLE. *
C      ******************************************************************
C
       DO 10 NEXTNR = 1, N
       IPRDCT = IPRDCT * NEXTNR
   10  CONTINUE
C
C      ******************************************************************
C      * STORE THE FACTORIAL OF N IN THE VARIABLE NFACT.                *
C      ******************************************************************
C
       NFACT = IPRDCT
C
C      ******************************************************************
C      * STEP 3.  READ IN THE K SETS OF NUMBERS OF OCCURRENCES OF OUT-   *
C      * COMES AND THEIR ASSOCIATED PROBABILITIES (X, P).  CALCULATE THE *
C      * REQUIRED FACTORIALS AND PRODUCTS.                              *
```

```
C     ****************************************************************
C
C
C     ****************************************************************
C     * INITIALIZE THE VARIABLES USED TO STORE THE PRODUCTS OF THE FACT-*
C     * ORIALS OF THE OCCURRENCES FOR EACH OUTCOME (PRODX) AND THE PROD-*
C     * UCTS OF THE PROBABILITIES (PRODPX).  THESE ARE INITIALIZED TO   *
C     * ONE SO THAT THE FIRST MULTIPLICATIONS INVOLVING THESE VARIABLES *
C     * WILL YIELD RESULTS EQUAL TO THE FIRST NUMBER BEING MULTIPLIED.  *
C     ****************************************************************
C
      PRODX = 1
      PRODPX = 1.0
C
C     ****************************************************************
C     * CYCLE THROUGH THE SET OF INSTRUCTIONS FROM THE ONE FOLLOWING THE*
C     * DO STATEMENT THROUGH AND INCLUDING THE ONE NUMBERED 40.  THE    *
C     * VARIABLE INDX WILL BE USED TO COUNT THE NUMBER OF CYCLES.  ITS  *
C     * INITIAL VALUE WILL BE ONE AND ITS FINAL VALUE WILL BE THE VALUE *
C     * OF K (THE NUMBER OF OUTCOMES).  SINCE THE COUNTER (INDX) WILL BE*
C     * INCREMENTED BY ONE AFTER EACH CYCLE, THE LOOP WILL BE EXECUTED  *
C     * K TIMES.                                                        *
C     ****************************************************************
C
      DO 40 INDX = 1, K
C
C     ****************************************************************
C     * READ THE NUMBER OF OCCURRENCES OF AN OUTCOME AND THE PROBABILITY*
C     * OF THE OUTCOME (X, P).                                          *
C     ****************************************************************
C
      READ (5, 501) X, P
  501 FORMAT (I6, F10.4)
C
C     ****************************************************************
C     * PRINT OUT THE NUMBER OF OCCURRENCES AND THE PROBABILITY OF THIS *
C     * OUTCOME.                                                        *
C     ****************************************************************
C
      WRITE (6, 601) INDX, P, X
  601 FORMAT (1X,8HOUTCOME , I6, 16H (PROBABILITY = , F10.3,
     *    12H) OCCURRING , I6, 6H TIMES)
C
C     ****************************************************************
C     * CALCULATE THE FACTORIAL OF X.                                  *
C     ****************************************************************
C
      IPRDCT = 1
      IF (X .EQ. 0) GO TO 30
      DO 20 NEXTNR = 1, X
      IPRDCT = IPRDCT * NEXTNR
   20 CONTINUE
C
C     ****************************************************************
C     * CALCULATE THE REQUIRED PRODUCTS.                              *
C     ****************************************************************
C
   30 PRODX = PRODX * IPRDCT
      PRODPX = PRODPX * (P ** X)
   40 CONTINUE
C
C     ****************************************************************
C     * STEP 4.  CALCULATE THE PROBABILITY (PROB).                    *
```

```
C     *******************************************************************
C
      PROB = (NFACT / PRODX) * PRODPX
C
C     *******************************************************************
C     * PRINT THE PROBABILITY.                                          *
C     *******************************************************************
C
      WRITE (6, 602) PROB
  602 FORMAT (1X, 3HIS , F10.3)
      STOP
      END
```

D. Example Problems

Example 1   Company XYZ employs production workers, salesmen, and management personnel. Sixty percent of the employment force are production workers, 25 percent are salesmen, and 15 percent are in management. A grievance committee of eight employees is to be formed.
What is the probability that of the 8 committee members 4 will be
production workers, 3 will be salesmen, and 1 will be management?

Input Data

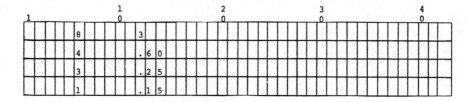

Solution   IN THE MULTINOMIAL EXPERIMENT WITH 8 TRIALS AND 3 OUT-
COMES, THE PROBABILITY OF
        OUTCOME 1 (PROBABILITY = 0.600) OCCURRING 4 TIMES
        OUTCOME 2 (PROBABILITY = 0.250) OCCURRING 3 TIMES
        OUTCOME 3 (PROBABILITY = 0.150) OCCURRING 1 TIMES
IS 0.085

Example 2   Find the probability of obtaining 2 fours, 1 five, and 3 sixes in five
rolls of a balanced die.

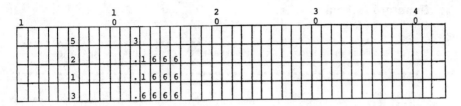

Input Data

IN THE MULTINOMIAL EXPERIMENT WITH 5 TRIALS AND 3 OUT- *Solution*
COMES, THE PROBABILITY OF
    OUTCOME 1 (PROBABILITY = 0.167) OCCURRING 2 TIMES
    OUTCOME 2 (PROBABILITY = 0.167) OCCURRING 1 TIMES
    OUTCOME 3 (PROBABILITY = 0.667) OCCURRING 3 TIMES
IS 0.014

☆ Poisson Distribution

A. Statement of Problem

The probability distribution of the Poisson random variable X, indicating the number of successes occurring in a given time interval, is given as:

$$P(X, \mu) = \frac{e^{-\mu}\mu^{x}}{x!}, \qquad x = 0, 1, 2, \ldots$$

where μ is the average number of successes occurring in a given time interval, and e is the base of the natural logarithms or 2.71828.
    Write a computer program to calculate the probability of the occurrence of a random variable X with the average number of successes occurring in a given time interval.

B. Algorithm

    1. Determine the average number of successes (MU). This value must be greater than zero.
    2. Repetitively execute the following steps with the values 0, 1, 2, ... for the Poisson random variable X. Continue executing these steps until the cumulative probability (CUMPRB) exceeds .9999.
        a. Evaluate the formula for the Poisson probability (PROB).
        b. Accumulate the Poisson probabilities.

C. General Program

```
C
C
C     **********************************************************************
C     * THIS PROGRAM COMPUTES THE PROBABILITY DISTRIBUTION AND THE CUMU-*
C     * LATIVE PROBABILITY DISTRIBUTION OF A POISSON RANDOM VARIABLE.    *
C     **********************************************************************
C
      REAL MU
      INTEGER X
C
C
C     **********************************************************************
C     * STEP 1.  DETERMINE THE AVERAGE NUMBER OF SUCCESSES (MU) AND IN- *
C     * SURE IT IS GREATER THAN ZERO.                                   *
C     **********************************************************************
C
      READ (5, 500) MU
  500 FORMAT (F10.3)
      IF (MU .LE. 0.0) STOP
C
C
C     **********************************************************************
C     * GENERATE THE FIRST TWO LINES OF PRINTED OUTPUT.                 *
C     **********************************************************************
C
      WRITE (6, 600) MU
  600 FORMAT (1X,35HTHE AVERAGE NUMBER OF SUCCESSES IS , F10.3)
      WRITE (6, 601)
  601 FORMAT (1X,40HVALUE OF X   PROBABILITY   CUM PROBABILITY)
C
C     **********************************************************************
C     * STEP 2   EVALUATE THE FORMULA FOR THE POISSON PROBABILITY FOR   *
C     * VALUES OF X OF 0, 1, 2, . . . AND ACCUMULATE THESE PROBABILITIES*
C     * USING THE VARIABLE CUMPRB.  CONTINUE UNTIL THE CUMULATIVE PROB- *
C     * ABILITY EXCEEDS .999.                                           *
C     **********************************************************************
C
C
C     **********************************************************************
C     * INITIALIZE THE VARIABLE USED TO ACCUMULATE THE PROBABILITIES    *
C     * (CUMPRB).                                                       *
C     **********************************************************************
C
      CUMPRB = 0.0
C
C     **********************************************************************
C     * INITIALIZE THE POISSON RANDOM VARIABLE (X) BY SETTING IT EQUAL  *
C     * TO ZERO.  BY REPETITIVELY ADDING ONE TO TO THIS VARIABLE WE WILL*
C     * GENERATE THE VALUES 0, 1, 2, . . . FOR THE RANDOM VARIABLE.     *
C     **********************************************************************
C
      X = 0
C
C     **********************************************************************
C     * STEP 2(A).  EVALUATE THE FORMULA FOR THE POISSON DISTRIBUTION.  *
C     * STORE THE PROBABILITY IN THE VARIABLE PROB.                     *
C     **********************************************************************
C
C
C     **********************************************************************
C     * CALCULATE THE FACTORIAL OF X.                                   *
C     **********************************************************************
C
    5 IPRDCT = 1
      IF (X .EQ. 0) GO TO 20
```

```
      DO 10 NEXTNR = 1, X
      IPRDCT = IPRDCT * NEXTNR
  10  CONTINUE
C
C     *************************************************************
C     * CALCULATE THE POISSON PROBABILITY (PROB).                 *
C     *************************************************************
C
  20  PROB = (EXP(-MU) * (MU ** X)) / IPRDCT
C
C     *************************************************************
C     * STEP 2(B).  ACCUMULATE THE PROBABILITIES.                 *
C     *************************************************************
C
      CUMPRB = CUMPRB + PROB
C
C     *************************************************************
C     * PRINT OUT THE VALUE OF THE POISSON RANDOM VARIABLE (X), THE *
C     * POISSON PROBABILITY (PROB) AND THE CUMULATIVE PROBABILITY    *
C     * (CUMPRB).                                                  *
C     *************************************************************
C
      WRITE (6, 602) X, PROB, CUMPRB
 602  FORMAT (3X, I6, 5X, F10.3, 5X, F10.3)
C
C     *************************************************************
C     * IF THE CUMULATIVE PROBABILITY (CUMPRB) IS GREATER THAN .999 THE *
C     * PROGRAM IS FINISHED.  OTHERWISE, CONTINUE TO THE NEXT INSTRUC-  *
C     * TION.                                                      *
C     *************************************************************
C
      IF (CUMPRB .GT. .999) GO TO 45
C
C     *************************************************************
C     * ADD ONE TO THE POISSON RANDOM VARIABLE (X).               *
C     *************************************************************
C
      X = X + 1
C
C     *************************************************************
C     * BRANCH BACK TO THE ROUTINE WHICH EVALUATES THE FORMULA FOR THE *
C     * POISSON PROBABILITY.                                       *
C     *************************************************************
C
      GO TO 5
  45  STOP
      END
```

D. Example Problems

At a busy intersection there are on the average two traffic accidents per week. What is the probability that during a given week there is exactly one traffic accident?

Example 1

Input Data

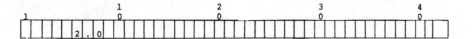

Solution   THE AVERAGE NUMBER OF SUCCESSES IS 2.000

VALUE OF X	PROBABILITY	CUM PROBABILITY
0	0.135	0.135
1	0.271	0.406
2	0.271	0.677
3	0.180	0.857
4	0.090	0.947
5	0.036	0.983
6	0.012	0.995
7	0.003	0.999
8	0.001	1.000

Example 2   During a given year a large urban city recorded homicides on the average of three per month. Find the probability that within a given month exactly five homicides are recorded.

Input Data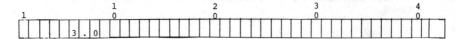

Solution   THE AVERAGE NUMBER OF SUCCESSES IS 3.000

VALUE OF X	PROBABILITY	CUM PROBABILITY
0	0.050	0.050
1	0.149	0.199
2	0.224	0.423
3	0.224	0.647
4	0.168	0.815
5	0.101	0.916
6	0.050	0.966
7	0.022	0.988
8	0.008	0.996
9	0.003	0.999
10	0.001	1.000

☆ Geometric Distribution

A. Statement of Problem

Suppose that repeated and independent trials of an experiment result in a success with a probability p and in failure with a probability $q =$

$1 - p$. Then the probability of the occurrence of the geometric random variable X, namely the number of the trial on which the first success occurs is found by the equation

$$g(x;\ p) = pq^{x-1}, x = 1, 2, 3 \ldots$$

Write a general computer program to compute the probability distribution of a geometric random variable.

B. Algorithm

1. Determine the probability of a success (P). Insure it is in the range 0 to 1.
2. Calculate the probability of a failure (Q) by subtracting the probability of a success (P) from 1.
3. Determine the probability distribution of the geometric random variable by repetitively evaluating the given formula with successive values of 1, 2,..., x until the probability is less than 0.001 and x is at least 20.

C. General Program

```
C
C    *************************************************************
C    * THIS PROGRAM COMPUTES THE PROBABILITY DISTRIBUTION OF A GEOMET- *
C    * RIC RANDOM VARIABLE.                                          *
C    *************************************************************
C
      INTEGER X
C
C    *************************************************************
C    * STEP 1.  DETERMINE THE PROBABILITY OF A SUCCESS (P) AND INSURE  *
C    * THAT IT IS EQUAL TO OR LESS THAN ONE AND EQUAL TO OR GREATER   *
C    * THAN ZERO.                                                    *
C    *************************************************************
C
      READ (5, 500) P
 500  FORMAT (F10.4)
      IF (P .LT. 0.0 .OR. P .GT. 1.0) STOP
C
C    *************************************************************
C    * STEP 2.  CALCULATE THE PROBABILITY OF A FAILURE (Q).          *
C    *************************************************************
C
      Q = 1.0 - P
C
C    *************************************************************
C    * STEP 3.  REPETITIVELY EVALUATE THE GIVEN FORMULA WITH VALUES OF *
C    * X OF 1, 2, . . .; CONTINUE UNTIL THE PROBABILITY IS LESS THAN  *
C    * .001 AND THE FORMULA HAS BEEN EVALUATED AT LEAST 20 TIMES.     *
C    *************************************************************
C
C
C    *************************************************************
C    * INITIALIZE THE GEOMETRIC RANDOM VARIABLE (X) BY SETTING IT EQUAL*
C    * TO ZERO.                                                      *
```

```
C       ********************************************************************
C
        X = 0
C
C       ********************************************************************
C       * ADD 1 TO THE GEOMETRIC RANDOM VARIABLE (X).  WE WILL GENERATE    *
C       * VALUES FOR X OF 1, 2, . . . BY REPETITIVELY ADDING ONE TO THIS   *
C       * VARIABLE.                                                        *
C       ********************************************************************
C
   25   X = X + 1
C
C       ********************************************************************
C       * EVALUATE THE GIVEN FORMULA; STORE THE PROBABILITY IN THE VARIA-  *
C       * BLE GPROB.                                                       *
C       ********************************************************************
C
        GPROB = P * (Q ** (X - 1))
C
C       ********************************************************************
C       * IF THE LIMITS HAVE BEEN REACHED ON THE PROBABILITY AND THE       *
C       * NUMBER OF EVALUATIONS STOP THE PROGRAM.                          *
C       ********************************************************************
C
        IF (GPROB .LT. .001 .AND. X .GT. 20) GO TO 40
C
C       ********************************************************************
C       * PRINT THE PROBABILITY.                                           *
C       ********************************************************************
C
        WRITE (6, 600) X, GPROB
  600   FORMAT (1X,56HTHE PROBABILITY OF THE FIRST SUCCESS OCCURRING ON TR
       *IAL , I6, 4H IS , F10.3)
C
C       ********************************************************************
C       * BRANCH BACK TO GENERATE THE NEXT VALUE OF X.                     *
C       ********************************************************************
C
        GO TO 25
   40   STOP
        END
```

D. Example Problems

Example 1   Find the probability that a person requires five tosses to obtain a
head from a balanced coin.

Input Data

Solution   THE PROBABILITY OF THE FIRST SUCCESS OCCURRING ON TRIAL
1 IS 0.500

THE PROBABILITY OF THE FIRST SUCCESS OCCURRING ON TRIAL
2 IS 0.250

THE PROBABILITY OF THE FIRST SUCCESS OCCURRING ON TRIAL
3 IS 0.125

THE PROBABILITY OF THE FIRST SUCCESS OCCURRING ON TRIAL
4 IS 0.063

THE PROBABILITY OF THE FIRST SUCCESS OCCURRING ON TRIAL
5 IS 0.031

THE PROBABILITY OF THE FIRST SUCCESS OCCURRING ON TRIAL
6 IS 0.016

THE PROBABILITY OF THE FIRST SUCCESS OCCURRING ON TRIAL
7 IS 0.008

THE PROBABILITY OF THE FIRST SUCCESS OCCURRING ON TRIAL
8 IS 0.004

THE PROBABILITY OF THE FIRST SUCCESS OCCURRING ON TRIAL
9 IS 0.002

THE PROBABILITY OF THE FIRST SUCCESS OCCURRING ON TRIAL
10 IS 0.001

THE PROBABILITY OF THE FIRST SUCCESS OCCURRING ON TRIAL
11 IS 0.000

THE PROBABILITY OF THE FIRST SUCCESS OCCURRING ON TRIAL
12 IS 0.000

THE PROBABILITY OF THE FIRST SUCCESS OCCURRING ON TRIAL
13 IS 0.000

THE PROBABILITY OF THE FIRST SUCCESS OCCURRING ON TRIAL.
14 IS 0.000

THE PROBABILITY OF THE FIRST SUCCESS OCCURRING ON TRIAL
15 IS 0.000

THE PROBABILITY OF THE FIRST SUCCESS OCCURRING ON TRIAL
16 IS 0.000

THE PROBABILITY OF THE FIRST SUCCESS OCCURRING ON TRIAL
17 IS 0.000

THE PROBABILITY OF THE FIRST SUCCESS OCCURRING ON TRIAL
18 IS 0.000

THE PROBABILITY OF THE FIRST SUCCESS OCCURRING ON TRIAL
19 IS 0.000

THE PROBABILITY OF THE FIRST SUCCESS OCCURRING ON TRIAL
20 IS 0.000

Example 2   Find the probability that a gambler requires four throws of a die to obtain either a 3 or a 6.

Input Data

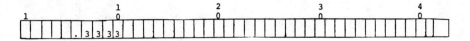

Solution   THE PROBABILITY OF THE FIRST SUCCESS OCCURRING ON TRIAL 1 IS 0.333

THE PROBABILITY OF THE FIRST SUCCESS OCCURRING ON TRIAL 2 IS 0.222

THE PROBABILITY OF THE FIRST SUCCESS OCCURRING ON TRIAL 3 IS 0.148

THE PROBABILITY OF THE FIRST SUCCESS OCCURRING ON TRIAL 4 IS 0.099

THE PROBABILITY OF THE FIRST SUCCESS OCCURRING ON TRIAL 5 IS 0.066

THE PROBABILITY OF THE FIRST SUCCESS OCCURRING ON TRIAL 6 IS 0.044

THE PROBABILITY OF THE FIRST SUCCESS OCCURRING ON TRIAL 7 IS 0.029

THE PROBABILITY OF THE FIRST SUCCESS OCCURRING ON TRIAL 8 IS 0.020

THE PROBABILITY OF THE FIRST SUCCESS OCCURRING ON TRIAL 9 IS 0.013

THE PROBABILITY OF THE FIRST SUCCESS OCCURRING ON TRIAL 10 IS 0.009

THE PROBABILITY OF THE FIRST SUCCESS OCCURRING ON TRIAL 11 IS 0.006

THE PROBABILITY OF THE FIRST SUCCESS OCCURRING ON TRIAL 12 IS 0.004

THE PROBABILITY OF THE FIRST SUCCESS OCCURRING ON TRIAL 13 IS 0.003

THE PROBABILITY OF THE FIRST SUCCESS OCCURRING ON TRIAL 14 IS 0.002

THE PROBABILITY OF THE FIRST SUCCESS OCCURRING ON TRIAL 15 IS 0.001

THE PROBABILITY OF THE FIRST SUCCESS OCCURRING ON TRIAL 16 IS 0.001

THE PROBABILITY OF THE FIRST SUCCESS OCCURRING ON TRIAL
17 IS 0.001

THE PROBABILITY OF THE FIRST SUCCESS OCCURRING ON TRIAL
18 IS 0.000

THE PROBABILITY OF THE FIRST SUCCESS OCCURRING ON TRIAL
19 IS 0.000

THE PROBABILITY OF THE FIRST SUCCESS OCCURRING ON TRIAL
20 IS 0.000

☆ Exercises

1. Three coins are tossed simultaneously. Find the cumulative probability distribution of obtaining 0, 1, 2, or 3 heads.

2. Given the following probability distribution:

$f(x_i)$	0.3	0.2	0.1	0.4
$X = x_i$	1	2	3	4

   find the cumulative probability distribution for obtaining 1, 2, 3, and 4 occurrences of the random variable X.

3. A multiple-choice examination given in General Psychology has 20 questions. Each question has four possible answers of which only one response is the correct answer. What is the probability that by sheer guesswork alone a student can obtain a total of exactly 15 correct answers?

4. A college student is enrolled in four courses during a given quarter. The student contends that he can obtain an A in a course with a probability of 0.2, a B with a probability of 0.3, and a C with a probability of 0.5 in each of his courses. Find the probability that during a given quarter the student obtains 2 Bs and 2 Cs.

5. A real estate agent sells on the average two homes per month. What is the probability that in any given month the agent will sell only one home? (Assume that monthly sales approximate a Poisson distribution.)

6. Given the probability function for the occurrence of the top face of a die as:

$f(x_i)$	⅙	⅙	⅙	⅙	⅙	⅙
x_i	1	2	3	4	5	6

   find the cumulative probability distribution.

Chapter 5

Estimation Theory

☆ **Estimating the Confidence Interval for a Mean of a Large Sample**
(σ known and $N \geq 30$)

A. Statement of Problem

Given the mean $\overline{X}$, standard deviation σ, sample size N of a sample, and a chosen level of confidence α, write a computer program to find a $(1 - \alpha)$ 100 percent confidence interval for μ, the mean of the population, based on the following relationship:

$$\overline{X} - Z_{\alpha/2} \cdot \frac{\sigma}{\sqrt{N}} < \mu < \overline{X} + Z_{\alpha/2} \cdot \frac{\sigma}{\sqrt{N}}$$

where $\overline{X}$ is the mean of a sample of size N taken from a normal population with known standard deviation σ and Z is the value of the standard normal random variable having an area of $\alpha/2$ to the right of the critical value $(Z_{\alpha/2})$.

B. Algorithm

1. Read in the population standard deviation (SIGMA), the sample mean (XBAR), the sample size (N), the confidence level (ALPHA), and the corresponding value of the standard normal random variable (ZALFA2). Insure that the sample size is equal to or greater than 30.

2. Compute the half interval (DISP) to the left or right of the mean by multiplying the value of the standard normal random variable by the quotient of the standard deviation (SIGMA) divided by the square root of the sample size (N).

3. Calculate the left (LOSIDE) of the confidence interval by subtracting the half interval (DISP) from the sample mean (XBAR). Calculate the right (HISIDE) side of the confidence interval by adding the half interval (DISP) to the sample mean (XBAR).

C. General Program

```
C     ******************************************************************
C     * THIS PROGRAM CALCULATES A CONFIDENCE INTERVAL FOR THE MEAN OF A *
C     * LARGE SAMPLE WHOSE MEAN IS KNOWN.  IT IS ASSUMED THAT THE STAND-*
C     * ARD DEVIATION OF THE POPULATION FROM WHICH THE SAMPLE WAS DRAWN *
C     * IS KNOWN.                                                       *
C     ******************************************************************
C
      INTEGER ALPHA
      REAL LOSIDE, HISIDE
C
C     ******************************************************************
C     * STEP 1.  READ THE VALUES FOR THE POPULATION STANDARD DEVIATION *
C     * (SIGMA), THE SAMPLE MEAN (XBAR), THE VALUE OF THE STANDARD NOR- *
C     * MAL RANDOM VARIABLE (ZALFA2), THE SAMPLE SIZE (N), AND THE CON- *
C     * FIDENCE LEVEL (ALPHA).                                         *
C     ******************************************************************
C
      READ (5, 500) SIGMA, XBAR, ZALFA2, N, ALPHA
  500 FORMAT (3F10.3, 2I6)
C
C     ******************************************************************
C     * STEP 2.  COMPUTE THE DISPLACEMENT FROM THE SAMPLE MEAN (DISP). *
C     ******************************************************************
C
      DISP = ZALFA2 * (SIGMA / N ** .5)
C
C     ******************************************************************
C     * STEP 3.  COMPUTE AND DISPLAY THE LEFT AND RIGHT SIDES OF THE   *
C     * CONFIDENCE INTERVAL.                                          *
C     ******************************************************************
C
      LOSIDE = XBAR - DISP
      HISIDE = XBAR + DISP
      WRITE (6, 600) SIGMA
  600 FORMAT (1X,37HTHE POPULATION STANDARD DEVIATION IS , F10.3)
      WRITE (6, 601) N
```

```
601  FORMAT (1X,19HTHE SAMPLE SIZE IS , I6)
     WRITE (6, 602) LOSIDE, HISIDE, ALPHA
602  FORMAT (1X,46HTHE CONFIDENCE INTERVAL FOR THE POPULATION IS ,
   * F10.3, 7H . . . , F10.3, 6H WITH, I6, 13H % CONFIDENCE)
     STOP
     END
```

D. Example Problems

The heights of a random sample of 50 college students showed a mean height of 67 inches with a standard deviation of 2.3 inches. Find a 95 percent confidence interval for the mean height of the entire college student population.

Example 1

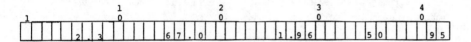

Input Data

THE POPULATION STANDARD DEVIATION IS 2.300

Solution

THE SAMPLE SIZE IS 50

THE CONFIDENCE INTERVAL FOR THE POPULATION IS 66.362 ... 67.638 WITH 95 PERCENT CONFIDENCE.

The mean and standard deviation for the grade point averages of a random sample of 40 business administration majors was calculated at 2.70 and 0.5, respectively. Find the 99 percent confidence interval for all business administration majors.

Example 2

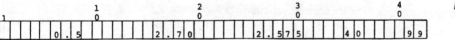

Input Data

THE POPULATION STANDARD DEVIATION IS 0.500

Solution

THE SAMPLE SIZE IS 40

THE CONFIDENCE INTERVAL FOR THE POPULATION IS 2.496 ... 2.904 WITH 99 PERCENT CONFIDENCE

☆ Estimating the Confidence Interval for a Mean of a Small Sample (σ unknown and $N < 30$)

A. Statement of Problem

Given a set of N variates $X_1, X_2, X_3, \ldots, X_N$, write a computer program to find a $(1 - \alpha)$ 100 percent confidence interval for μ, the mean of the population, based on the following relationship:

$$\overline{X} - t_{\alpha/2} \cdot \frac{s}{\sqrt{N}} < \mu < \overline{X} + t_{\alpha/2} \cdot \frac{s}{\sqrt{N}}$$

where the mean $\overline{X}$ and standard deviation s are to be computed from the N variates, and $t_{\alpha/2}$ is the value of the t distribution with $\nu = n - 1$ degrees of freedom having an area of $\alpha/2$ under the distribution and to the right of the critical value $(t_{\alpha/2})$.

B. Algorithm

1. Determine the number (N) of data points in the sample, the confidence level (ALPHA) and the corresponding value of the T distribution (TALFA2).

2. Calculate the standard deviation of the sample (SSIGMA) by evaluating the formula:

$$\text{SSIGMA} = \sqrt{\frac{N \sum X^2 - (\sum X)^2}{N \cdot (N - 1)}}$$

Compute the mean of the sample (XBAR).

3. Compute the half interval (DISP) to the right or left of the mean by multiplying the value of the T distribution by the sample standard deviation (SSIGMA) divided by the square root of the sample size (N). Then calculate the left side (LOSIDE) of the confidence interval by subtracting the half interval (DISP) from the sample mean (XBAR). The right side (HISIDE) is calculated by adding the half interval to the sample mean (XBAR).

C. General Program

```
C
C     **********************************************************************
C     * THIS PROGRAM CALCULATES A CONFIDENCE INTERVAL FOR THE POPULATION*
C     * MEAN BASED UPON A SMALL SAMPLE.  THE POPULATION STANDARD DEVIA- *
C     * TION IS NOT KNOWN.                                              *
```

```
C     *****************************************************************
C
      INTEGER ALPHA
      REAL LOSIDE, HISIDE
C
C     *****************************************************************
C     * STEP 1.  DETERMINE THE NUMBER OF DATA POINTS IN THE SAMPLE (N), *
C     * THE DESIRED VALUE OF THE T DISTRIBUTION (TALFA2), AND THE CONFI-*
C     * DENCE LEVEL (ALPHA).                                           *
C     *****************************************************************
C
      READ (5, 500) N, TALFA2, ALPHA
  500 FORMAT (I6, F10.3, I6)
C
C     *****************************************************************
C     * STEP 2.  CALCULATE THE STANDARD DEVIATION OF THE SAMPLE; ACCUM- *
C     * ULATE THE SUM OF THE SAMPLE DATA POINTS (SUMX) AS WELL AS THE   *
C     * SUM OF THEIR SQUARED VALUES (SUMXSQ).                          *
C     *****************************************************************
C
C
C     *****************************************************************
C     * INITIALIZE THE ACCUMULATORS OF THE SUM OF THE DATA POINTS AND   *
C     * THE SUM OF THE SQUARES OF THE DATA POINTS.                     *
C     *****************************************************************
C
      SUMX = 0.0
      SUMXSQ = 0.0
C
C     *****************************************************************
C     * REPETITIVELY CYCLE THROUGH THE SET OF INSTRUCTIONS FROM THE ONE *
C     * FOLLOWING THE DO STATEMENT THROUGH AND INCLUDING THE STATEMENT  *
C     * NUMBERED 10.  CYCLE THROUGH THESE INSTRUCTIONS N TIMES.        *
C     *****************************************************************
C
      DO 10 I = 1, N
      READ (5, 501) X
  501 FORMAT (F10.3)
      SUMX = SUMX + X
   10 SUMXSQ = SUMXSQ + (X * X)
C
C     *****************************************************************
C     * CALCULATE THE SAMPLE MEAN (XBAR).                             *
C     *****************************************************************
C
      XBAR = SUMX / N
C
C     *****************************************************************
C     * CALCUALTE THE SAMPLE STANDARD DEVIATION (SSIGMA).            *
C     *****************************************************************
C
      SSIGMA = SQRT(((N * SUMXSQ) - (SUMX ** 2)) / (N * (N - 1)))
C
C     *****************************************************************
C     * STEP 3.  CALCULATE THE REQUIRED CONFIDENCE INTERVAL.         *
C     *****************************************************************
C
C
C     *****************************************************************
C     * CALCULATE THE DISPLACEMENT FROM THE SAMPLE MEAN (DISP).      *
C     *****************************************************************
C
      DISP = TALFA2 * (SSIGMA / N ** .5)
```

123

```
C
C
C    ****************************************************************
C    * CALCULATE THE LEFT AND RIGHT SIDES OF THE CONFIDENCE INTERVAL.  *
C    ****************************************************************
C
     LOSIDE = XBAR - DISP
     HISIDE = XBAR + DISP
     WRITE (6, 601) N
 601 FORMAT (1X,19HTHE SAMPLE SIZE IS , I6)
     WRITE (6, 602) SSIGMA
 602 FORMAT (1X,33HTHE SAMPLE STANDARD DEVIATION IS , F10.3)
     WRITE (6, 603) XBAR
 603 FORMAT (1X,19HTHE SAMPLE MEAN IS , F10.3)
     WRITE (6, 604) LOSIDE , HISIDE, ALPHA
 604 FORMAT (1X,27HTHE CONFIDENCE INTERVAL IS , F10.3, 7H . . . ,
    *    F10.3, 6H WITH , I6, 13H % CONFIDENCE)
     STOP
     END
```

D. Example Problems

Example 1   A sample of 10 units and their diameters of a 10,000-unit shipment of bolts yielded the following sample diameters (in cm): 10.1, 9.3, 9.7, 9.1, 10.2, 8.1, 10.7, 10.1, 10.2, and 9.9. Calculate a 95 percent confidence interval of the diameters for the entire shipment of bolts.

Input Data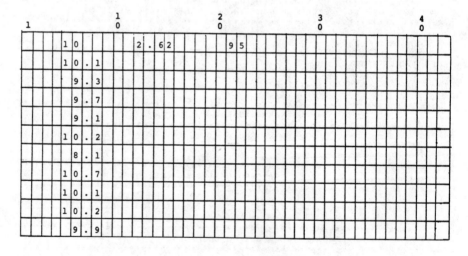

Solution   THE SAMPLE SIZE IS 10

THE SAMPLE STANDARD DEVIATION IS 0.740

THE SAMPLE MEAN IS 9.740

THE CONFIDENCE INTERVAL IS 9.127 ... 10.353 WITH 95 PERCENT   *Example 2*
CONFIDENCE

Find the 99 percent confidence interval for the average GPA (grade
point average) for a senior high school class based upon the following
sample GPA's of twenty junior students:

2.37	2.01	3.42	2.09
2.93	1.96	2.19	3.36
3.72	1.47	2.47	3.21
2.57	2.30	2.37	1.13
2.43	1.73	2.50	2.65

Input Data

Scale markers: 1   1 0   2 0   3 0   4 0

Grid entries:

```
2 0          2. 8 6 1        9 9
2 . 3 7
2 . 9 3
3 . 7 2
2 . 5 7
2 . 4 3
2 . 0 1
1 . 9 6
1 . 4 7
2 . 3 0
1 . 7 3
3 . 4 2
2 . 1 9
2 . 4 7
2 . 3 7
2 . 5 0
2 . 0 9
3 . 3 6
3 . 2 1
1 . 1 3
2 . 6 5
```

Solution   THE SAMPLE SIZE IS 20

THE SAMPLE STANDARD DEVIATION IS 0.654

THE SAMPLE MEAN IS 2.444

THE CONFIDENCE INTERVAL IS 2.025 ... 2.863 WITH 99 PERCENT CONFIDENCE

☆ **Estimating the Confidence Interval for the Difference Between Two Means (σ_1^2 and σ_2^2 are known; n_1, $n_2 \geqslant 30$)**

A. Statement of Problem

Given the means ($\overline{X}_1$ and $\overline{X}_2$), variances (σ_1^2 and σ_2^2) and the number of variates (n_1 and n_2) of two independent populations, write a computer program to find a $(1 - \alpha)$ 100 percent confidence interval for $\mu_1 - \mu_2$, the difference between the means of two populations, based upon the following relationship:

$$(\overline{X}_1 - \overline{X}_2) - Z_{\alpha/2} \sqrt{\frac{\sigma_1^2}{n_1} + \frac{\sigma_2^2}{n_2}} < \mu_1 - \mu_2 < (\overline{X}_1 - \overline{X}_2) + Z_{\alpha/2} \sqrt{\frac{\sigma_1^2}{n_1} + \frac{\sigma_2^2}{n_2}}$$

where Z is the value of the standard normal random variable having an area of $\alpha/2$ to the right of the critical value ($Z_{\alpha/2}$).

B. Algorithm

1. Determine the sample sizes (N1, N2), the corresponding sample means (XBAR1, XBAR2) and variances (VAR1, VAR2), the confidence level (ALPHA) and the corresponding value of the standard normal random variable (ZALFA2). Insure that the sample sizes are equal to or greater than 30.

2. Calculate the half-interval (DISP) to the left or right of the estimated difference between the two means by multiplying the value of the standard normal random variable by the square root of the sum of the variance of the first sample (VAR1) divided by its sample size (N1) plus the variance of the second sample (VAR2) divided by its sample size (N2).

3. Calculate the left side (LOSIDE) of the confidence interval by subtracting the half-interval (DISP) from the absolute value of the difference between the sample means (XBAR1, XBAR2). Calculate the right side (HISIDE) of the confidence interval by adding the half-interval (DISP) to the absolute

value of the difference between the sample means (XBAR1, XBAR2).

C. General Program

```
C
C     ********************************************************************
C     * THIS PROGRAM CALCULATES A CONFIDENCE INTERVAL FOR THE DIFFERENCE*
C     * BETWEEN TWO MEANS.  THE CALCULATION IS BASED UPON TWO LARGE     *
C     * SAMPLES WHOSE MEANS AND VARIANCES ARE KNOWN.                    *
C     ********************************************************************
C
      INTEGER ALPHA
      REAL LOSIDE, HISIDE
C
C     ********************************************************************
C     * STEP 1.  DETERMINE THE SAMPLE SIZES (N1, N2), THE CORRESPONDING *
C     * SAMPLE MEANS (XBAR1, XBAR2), THE CORRESPONDING VARIANCES (VAR1, *
C     * VAR2), THE VALUE OF THE STANDARD NORMAL RANDOM VARIABLE (ZALFA2)*
C     * AND THE CONFIDENCE LEVEL (ALPHA).  INSURE THAT THE SAMPLE SIZES *
C     * ARE GREATER THAN OR EQUAL TO 30.                                *
C     ********************************************************************
C
      READ (5, 500) N1, N2, XBAR1, XBAR2, VAR1, VAR2, ZALFA2, ALPHA
  500 FORMAT (2I6, 5F10.3, I6)
      IF (N1 .LT. 30 .OR. N2 .LT. 30) STOP
C
C     ********************************************************************
C     * STEP 2.  CALCULATE THE DISPLACEMENT (DISP) ON EITHER SIDE OF THE*
C     * DIFFERENCE BETWEEN MEANS.                                       *
C     ********************************************************************
C
      DISP = ZALFA2 * SQRT ((VAR1 / N1) + (VAR2 / N2))
C
C     ********************************************************************
C     * STEP 3.  CALCULATE AND DISPLAY THE CONFIDENCE INTERVAL.         *
C     ********************************************************************
C
      LOSIDE = ABS (XBAR1 - XBAR2) - DISP
      HISIDE = ABS(XBAR1 - XBAR2) + DISP
      WRITE (6, 600) N1, N2
  600 FORMAT (1X,21HTHE SAMPLE SIZES ARE , 2I6)
      WRITE (6, 601) XBAR1, XBAR2
  601 FORMAT (1X,21HTHE SAMPLE MEANS ARE, 2F10.3)
      WRITE (6, 602) VAR1, VAR2
  602 FORMAT (1X,25HTHE SAMPLE VARIANCES ARE , 2F10.3)
      WRITE (6, 603)
  603 FORMAT (1X,56HTHE CONFIDENCE INTERVAL FOR THE DIFFERENCE BETWEEN M
     *EANS)
      WRITE (6, 604) LOSIDE, HISIDE, ALPHA
  604 FORMAT (1X, 5X, F10.3, 7H . . . , F10.3, 6H WITH , I6,
     *   13H % CONFIDENCE)
      STOP
      END
```

D. Example Problems

Example 1

A physics laboratory examination was given to one section of 60 boys and 40 girls. The boys had an average grade of 82 with a standard deviation of 5, while the girls had an average grade of 74 with a standard deviation of 9. Find a 99 percent confidence interval for the difference between the mean scores of all boys and girls taking the examination.

Input Data

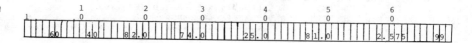

Solution   THE SAMPLE SIZES ARE 60   40

THE SAMPLE MEANS ARE 82.000   74.000

THE SAMPLE VARIANCES ARE 25.000   81.000

THE CONFIDENCE INTERVAL FOR THE DIFFERENCE BETWEEN MEANS 3.976 ... 12.024 WITH 99 PERCENT CONFIDENCE

Example 2   Two manufacturers of television picture tubes ran a competitive study. Manufacturer A claimed that the average life expectancy of their company's picture tubes was equivalent to those produced by manufacturer B. A random sample of 36 picture tubes from manufacturer A showed the average life expectancy to be 4 years with a standard deviation of 5, while a random sample of 45 picture tubes taken from manufacturer B showed their life expectancy to be 5 years with a standard deviation of 1.5 years. Find a 99 percent confidence interval for the differences between means for the two companies.

Input Data

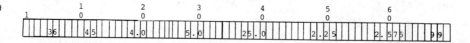

Solution   THE SAMPLE SIZES ARE 36   45

THE SAMPLE MEANS ARE 4.000   5.000

THE SAMPLE VARIANCES ARE 25.000   2.250

THE CONFIDENCE INTERVAL FOR THE DIFFERENCE BETWEEN MEANS −1.222 ... 3.222 WITH 99 PERCENT CONFIDENCE

☆ Estimating the Confidence Interval for the Difference Between Two Means
$$(\sigma_1^2 = \sigma_2^2;\ n_1,\ n_2 < 30)$$

A. Statement of Problem

Given a set of n_1 variates $X_1, X_2, \ldots, X_N$ and a set of n_2 variates $Y_1, Y_2, \ldots, Y_{n_2}$, write a computer program to find a $(1 - \alpha)$ 100 percent confidence interval for $\mu - \mu_2$, the difference between the means of two populations, based upon the following relationship:

$$(\overline{X}_1 - \overline{X}_2) - t_{\alpha/2}S_p \sqrt{\frac{1}{n_1} + \frac{1}{n_2}} < \mu_1 - \mu_2 < (\overline{X}_1 - \overline{X}_2) + t_{\alpha/2}S_p \sqrt{\frac{1}{n_1} + \frac{1}{n_2}}$$

$$\text{where } S_p = \sqrt{\frac{(n_1 - 1)S_1^2 + (n_2 - 1)S_2^2}{n_1 + n_2 - 2}}$$

and $\overline{X}_1$ and $\overline{X}_2$ are the means of two independent random samples from populations with approximately equivalent variances $\sigma_1^2 = \sigma_2^2$; Z is the value of the standard normal random variable having an area of $\alpha/2$ to the right of the critical value $(t_{\alpha/2})$.

B. Algorithm

1. There are two options. Option 1 assumes that the sample means (XBAR1, XBAR2) and the variances (VAR1, VAR2) are known and included as input. Option 2 assumes that the input consists of raw data. Determine the value of the option (OPTION). If the value of the option is 2, proceed to Step 3; otherwise, continue to Step 2.

2. Determine the sample sizes (N_1, N_2), the sample means (XBAR1, XBAR2), and the sample variances (VAR1, VAR2). Then proceed to Step 4.

3. Read each sample and calculate their means (XBAR1, XBAR2) and the variances (VAR1, VAR2).

4. Calculate the pooled variance (PVAR) as indicated in the given formula.

5. Determine the confidence level (ALPHA) and the corresponding value of the T distribution (TALFA2).

6. Compute the half-interval (DISP) to the left or right of the estimated difference between the population means by multiplying the value of the T distribution (TALFA2) by the pooled

variance (PVAR) by the square root of the sum of the reciprocals of the sample sizes. Calculate the left side (LOSIDE) of the confidence interval by subtracting the half-interval (DISP) from the difference between the sample means. Calculate the right side (HISIDE) of the confidence interval by adding the half-interval (DISP) to the difference between the sample means.

C. General Program

```
C
C
C     ***********************************************************************
C     * THIS PROGRAM CALCULATES A CONFIDENCE INTERVAL FOR THE DIFFERENCE*
C     * BETWEEN POPULATION MEANS BASED UPON SMALL SAMPLES.  IT ASSUMES  *
C     * THAT THE POPULATION VARIANCES ARE EQUAL.                        *
C     ***********************************************************************
C
      INTEGER OPTION, ALPHA
      REAL LOSIDE, HISIDE
C
C
C     ***********************************************************************
C     * STEP 1.  DETERMINE THE VALUE OF THE OPTION.  IF THIS VALUE IS 1 *
C     * THE PROGRAM ASSUMES THAT THE SAMPLE MEANS AND VARIANCES ARE     *
C     * KNOWN AND GIVEN AS INPUT.  IF THE VALUE IS 2 THE PROGRAM ASSUMES*
C     * THAT THE INPUT CONSISTS OF RAW DATA.                            *
C     ***********************************************************************
C
      READ (5, 500) OPTION
  500 FORMAT (I6)
C
C
C     ***********************************************************************
C     * IF THE VALUE OF OPTION IS 1 BRANCH TO THE FIRST OF THE STATE-   *
C     * MENT NUMBERS APPEARING INSIDE THE PARENTHESES OF THE FOLLOWING  *
C     * INSTRUCTION.  IF THE VALUE OF OPTION IS 2 BRANCH TO THE SECOND  *
C     * STATEMENT NUMBER APPEARING INSIDE THE PARENTHESES.  IF THE VALUE*
C     * OF OPTION IS NEITHER A 1 OR A 2 THE PROGRAM BRANCHES TO NEITHER *
C     * OF THE GIVEN STATEMENT NUMBERS BUT CONTINUES WITH THE NEXT IN-  *
C     * STRUCTION.                                                      *
C     ***********************************************************************
C
      GO TO (10, 20), OPTION
      STOP
C
C
C     ***********************************************************************
C     * STEP 2.  THIS IS OPTION 1 AND ASSUMES THAT THE SAMPLE SIZES,    *
C     * MEANS AND VARIANCES ARE GIVEN.  READ THESE IN AND THEN BRANCH TO*
C     * STEP 4.                                                         *
C     ***********************************************************************
C
   10 READ (5, 504) N1, N2, XBAR1, XBAR2, VAR1, VAR2
  504 FORMAT (2I6,F7.0,F10.2,2X,F10.2,F8.0)
      GO TO 40
C
C
C     ***********************************************************************
C     * STEP 3.  THIS IS OPTION 2.  READ IN THE TWO SAMPLES AND CALCU-  *
C     * LATE THEIR MEANS AND VARIANCES.                                 *
C     * LOOP THROUGH STEP 3 TWICE (BECAUSE THERE ARE 2 SAMPLES).        *
C     ***********************************************************************
C
   20 DO 30 I = 1, 2
```

```
C
C     ****************************************************************
C     * INITIALIZE THE ACCUMULATORS OF THE SUM OF THE DATA POINTS (SUMX)*
C     * AND THE SUM OF THE SQUARES OF THE DATA POINTS (SUMXSQ).          *
C     ****************************************************************
C
      SUMX = 0.0
      SUMXSQ = 0.0
C
C     ****************************************************************
C     * READ IN THE NEXT SAMPLE SIZE.                                   *
C     ****************************************************************
C
      READ (5, 501) N
  501 FORMAT (I6)
C
C     ****************************************************************
C     * READ IN THE SAMPLE DATA POINTS AND ACCUMULATE THEIR SUM AND THE *
C     * SUM OF THEIR SQUARES.                                           *
C     ****************************************************************
C
      DO 15 J = 1, N
      READ (5, 502) X
  502 FORMAT (F10.3)
      SUMX = SUMX + X
   15 SUMXSQ = SUMXSQ + (X * X)
C
C     ****************************************************************
C     * CALCULATE THE VARIANCE AND THE MEAN FOR THIS SAMPLE.            *
C     ****************************************************************
C
      VAR = ((N * SUMXSQ) - (SUMX * SUMX)) / (N * (N - 1))
      XBAR = SUMX / N
C
C     ****************************************************************
C     * IF THIS IS THE FIRST SAMPLE STORE THE MEAN, VARIANCE AND SAMPLE *
C     * SIZE IN XBAR1, VAR1 AND N1 RESPECTIVELY.  OTHERWISE, STORE THEM *
C     * IN XBAR2, VAR2 AND N2 RESPECTIVELY.                             *
C     ****************************************************************
C
      GO TO (50, 60), I
   50 XBAR1 = XBAR
      VAR1 = VAR
      N1 = N
      GO TO 30
   60 XBAR2 = XBAR
      VAR2 = VAR
      N2 = N
   30 CONTINUE
C
C     ****************************************************************
C     * STEP 4.  CALCULATE THE POOLED VARIANCE (PVAR).                  *
C     ****************************************************************
C
   40 PVAR = (((N1 - 1) * VAR1) + ((N2 - 1) * VAR2)) / (N1 + N2 - 2)
C
C     ****************************************************************
C     * STEP 5.  DETERMINE THE VALUE OF T (TALFA2) AND THE CONFIDENCE   *
C     * LEVEL (ALPHA).                                                  *
C     ****************************************************************
C
      READ (5, 503) TALFA2, ALPHA
  503 FORMAT (F10.2, I6)
```

```
      C
      C
      C     **********************************************************************
      C     * STEP 6.  CALCULATE THE DISPLACEMENT (DISP) ON EITHER SIDE OF THE*
      C     * DIFFERENCE BETWEEN THE POPULATION MEANS.                        *
      C     **********************************************************************
      C
            DISP = TALFA2 * SQRT(PVAR) * SQRT((1.0 / N1) + (1.0 / N2))
      C
      C     **********************************************************************
      C     * CALCULATE AND PRINT THE CONFIDENCE INTERVAL.                    *
      C     **********************************************************************
      C
            LOSIDE = (XBAR1 - XBAR2) - DISP
            HISIDE = (XBAR1 - XBAR2) + DISP
            WRITE (6, 600) N1, N2
      600   FORMAT (1X,21HTHE SAMPLE SIZES ARE , 2I6)
            WRITE (6, 601) XBAR1, XBAR2
      601   FORMAT (1X,21HTHE SAMPLE MEANS ARE , 2F10.3)
            WRITE (6, 602) VAR1, VAR2
      602   FORMAT (1X,25HTHE SAMPLE VARIANCES ARE , 2F10.3)
            WRITE (6, 603) PVAR
      603   FORMAT (1X,23HTHE POOLED VARIANCE IS , F10.3)
            WRITE (6, 604)
      604   FORMAT (1X,56HTHE CONFIDENCE INTERVAL FOR THE DIFFERENCE BETWEEN M
           *EANS)
            WRITE (6, 605) LOSIDE, HISIDE, ALPHA
      605   FORMAT (1X, 6X, F10.3, 7H . . . , F10.3, 6H WITH , I6,
           *   13H % CONFIDENCE)
            STOP
            END
```

D. Example Problems

Example 1   Two groups of students were given separate methods of statistics instruction, lecture and TV. The final averages in the statistics course for each of two instructional methods are shown as follows for 22 students:

Lecture	TV Lecture
87	69
97	92
83	79
84	86
89	82
74	67
62	69
52	54
73	70
77	68
85	72
93	77
84	67
82	71
78	67
77	68
79	70

Lecture	TV Lecture
62	55
84	73
69	61
70	63
74	60

Calculate a 95 percent confidence interval for the difference in means between the two methods.

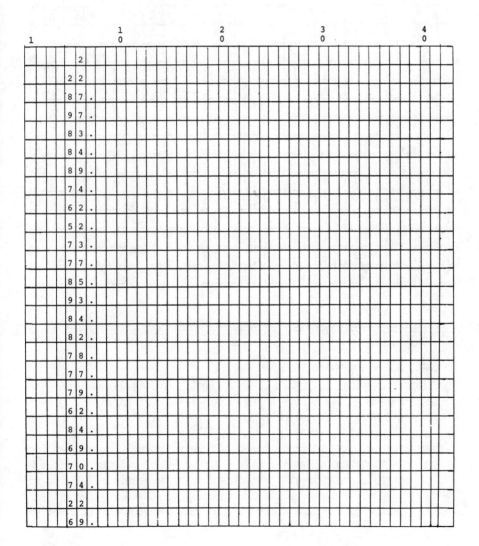

Input Data

```
1              1              2              3              4
               0              0              0              0

    9 2 .
    7 9 .
    8 6 .
    8 2 .
    6 7 .
    6 9 .
    5 4 .
    7 0 .
    6 8 .
    7 2 .
    7 7 .
    6 7 .
    7 1 .
    6 7 .
    6 8 .
    7 0 .
    5 5 .
    7 3 .
    6 1 .
    6 3 .
    6 0 .
    2 . 0 2 1       9 5
```

Solution   THE SAMPLE SIZES ARE 22   22

THE SAMPLE MEANS ARE 77.955   70.000

THE SAMPLE VARIANCES ARE 113.284   84.571

THE POOLED VARIANCE IS 98.927

THE CONFIDENCE INTERVAL FOR THE DIFFERENCE BETWEEN MEANS 1.894 ... 14.015 WITH 95 PERCENT CONFIDENCE

Example 2   Two groups of women were given different types of exercises for a weight reduction program. Group A contained 12 women who lost, on the average, eight pounds with a standard deviation of 0.5 pound over a two-month period; Group B contained 20 women who lost, on the average, four pounds with a standard deviation of 2.0 pounds

over the same period of time. Find a 95 percent confidence interval
for the differences in mean weight loss between the two groups.

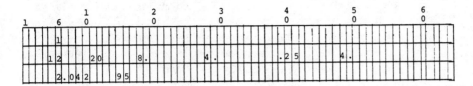

Input Data

THE SAMPLE SIZES ARE 12   20   *Solution*

THE SAMPLE MEANS ARE 8.000   4.000

THE SAMPLE VARIANCES ARE 0.250   4.000

THE POOLED VARIANCE IS 2.625

THE CONFIDENCE INTERVAL FOR THE DIFFERENCE BETWEEN
MEANS 2.792 ... 5.208 WITH 95 PERCENT CONFIDENCE

☆ Estimating the Confidence Interval for a Proportion p ($N \geqslant 30$)

A. Statement of Problem

Given a proportion ($\hat{p}$) of x successes in a random sample of size n,
where $p = x/n$, write a computer program to find the confidence
interval for the binomial parameter p as given by the equation:

$$\hat{p} - Z_{\alpha/2}\sqrt{\frac{\hat{p}\hat{q}}{n}} < p < \hat{p} + Z_{\alpha/2}\sqrt{\frac{\hat{p}\hat{q}}{n}}$$

where $\hat{q} = 1 - \hat{p}$ and Z is the value of the standard normal random
variable having an area of $\alpha/2$ to the right of the critical value ($Z_{\alpha/2}$).

B. Algorithm

1. Determine the sample size (N), the number of successes in
   the sample (X), the confidence interval (ALPHA), and the
   corresponding value of the standard normal random variable
   (ZALFA2). Insure that the sample size is equal to or greater
   than 30 and that the number of successes do not exceed
   the sample size.

2. Calculate the proportion of successes (PHAT) by dividing the

number of successes (X) by the sample size (N). Calculate the proportion of failures (QHAT) by subtracting the proportion of successes (PHAT) from 1.0.

3. Calculate the half-interval (DISP) to the left or right of the estimate population proportion (PHAT) by multiplying the value of the standard normal random variable (ZALFA2) by the square root of the proportion of successes (PHAT) times the proportion of failures (QHAT) divided by the sample size (N).

4. Calculate the left (LOSIDE) side of the confidence interval by subtracting the half-interval (DISP) from the estimated population proportion (PHAT). Calculate the right (HISIDE) side of the confidence interval by adding the half-interval (DISP) to the estimated population proportion (PHAT).

C. General Program

```
C
C
C     *********************************************************************
C     * THIS PROGRAM COMPUTES A CONFIDENCE INTERVAL FOR A PROPORTION      *
C     * OF A POPULATION BASED UPON A LARGE SAMPLE.                        *
C     *********************************************************************
C
      INTEGER X, ALPHA
      REAL LOSIDE, HISIDE
C
C     *********************************************************************
C     * STEP 1.  DETERMINE THE SAMPLE SIZE (N), THE NUMBER OF SUCCESSES *
C     * IN THE SAMPLE (X), THE VALUE OF THE STANDARD NORMAL RANDOM VARI-*
C     * ABLE (ZALFA2) AND THE CONFIDENCE LEVEL (ALPHA).  INSURE THAT THE*
C     * NUMBER OF SUCCESSES IS EQUAL TO OR LESS THAN THE SAMPLE SIZE AND*
C     * THAT THE SAMPLE SIZE IS EQUAL TO OR GREATER THAN 30.            *
C     *********************************************************************
C
      READ (5, 500) N, X, ZALFA2, ALPHA
  500 FORMAT (2I6, F10.3, I6)
      IF (N .LT. 30) STOP
      IF (X .GT. N) STOP
C
C     *********************************************************************
C     * STEP 2.  CALCULATE THE PROPORTION OF SUCCESSES (PHAT) AND THE    *
C     * PROPORTION OF FAILURES (QHAT).  THE FUNCTION FLOAT MERELY CON-   *
C     * VERTS THE INTEGER VALUE OF N TO A REAL VALUE; THIS RESULTANT     *
C     * REAL VALUE IS USED IN THE COMPUTATIONS.                          *
C     *********************************************************************
C
      PHAT = X / FLOAT (N)
      QHAT = 1.0 - PHAT
C
C     *********************************************************************
C     * STEP 3.  CALCULATE THE DISPLACEMENT (DISP) ON EITHER SIDE OF THE*
C     * POPULATION PROPORTION.                                          *
C     *********************************************************************
C
      DISP = ZALFA2 * SQRT ((PHAT * QHAT) / N)
```

```
C
C     **********************************************************************
C     *  STEP 4.   CALCULATE AND PRINT THE CONFIDENCE INTERVAL.           *
C     **********************************************************************
C
      LOSIDE = PHAT - DISP
      HISIDE = PHAT + DISP
      WRITE (6, 600) N
 600  FORMAT (1X,19HTHE SAMPLE SIZE IS , I6)
      WRITE (6, 601) PHAT
 601  FORMAT (1X,31HTHE PROPORTION OF SUCCESSES IS , F10.3)
      WRITE (6, 602) LOSIDE, HISIDE, ALPHA
 602  FORMAT (1X,46HTHE CONFIDENCE INTERVAL FOR THE PROPORTION IS ,
     *    F10.3, 7H . . . , F10.3, 6H WITH , I6, 13H % CONFIDENCE)
      STOP
      END
```

D. Example Problems

A sample poll of 1000 citizens chosen at random from all voters indicated that 557 favored the incumbent Republican, whereas the rest favored the Democratic challenger. Find a 95 percent confidence interval for the actual proportion of voters favoring the Republican.

Example 1

Input Data

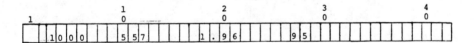

THE SAMPLE SIZE IS 1000

THE PROPORTION OF SUCCESSES IS 0.557

THE CONFIDENCE INTERVAL FOR THE PROPORTION IS 0.526 ...
0.588 WITH 95 PERCENT CONFIDENCE.

Solution

A random sample of machinery in a large factory showed that 4 machines were nonfunctional while 28 were operational. Find a 99 percent confidence interval for the actual proportion of nonfunctional machines in the entire factory.

Example 2

Input Data

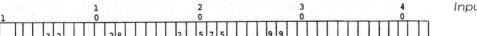

Solution   THE SAMPLE SIZE IS 32

THE PROPORTION OF SUCCESSES IS 0.875

THE CONFIDENCE INTERVAL FOR THE PROPORTION IS 0.724 ...
1.026 WITH 99 PERCENT CONFIDENCE.

☆ Estimating the Confidence Interval for the Difference Between the Two Proportions (n_1, $n_2 \geqslant$ 30)

A. Statement of Problem

Given the proportion of successes of $\hat{p}_1$ of a random sample size n_1
and the proportion of successes $\hat{p}_2$ of a random sample of size n_2,
write a computer program to find the confidence interval for
$\hat{p}_1 - \hat{p}_2$, the difference of two binomial parameters, as given by the
following equation:

$$(\hat{p}_1 - \hat{p}_2) - Z_{\alpha/2} \sqrt{\frac{\hat{p}_1 \hat{q}_1}{n_1} + \frac{\hat{p}_2 \hat{q}_2}{n_2}} < p_1 - p_2 < (\hat{p}_1 - \hat{p}_2)$$

$$+ Z_{\alpha/2} \sqrt{\frac{\hat{p}_1 \hat{q}_1}{n_1} + \frac{\hat{p}_2 \hat{q}_2}{n_2}}$$

where $\hat{q}_1 = 1 - \hat{p}_1$ and $\hat{q}_2 = 1 - \hat{p}_2$; Z is the value of the standard
normal random variable having an area of $\alpha/2$ to the right of the critical
value ($Z_{\alpha/2}$).

B. Algorithm

1. Determine the sample sizes (N1, N2) the corresponding
   number of successes (X1, X2), the confidence level (ALPHA),
   and the corresponding value of the standard normal random
   variable (ZALFA2). Insure that the sample sizes are equal to
   or greater than 30 and that the number of successes does
   not exceed the corresponding sample sizes.

2. Calculate the proportion of successes in each sample (PHAT1,
   PHAT2) by dividing the number of successes (X1, X2) by the
   corresponding sample sizes (N1, N2). Compute the proportion
   of failures in each group (QHAT1, QHAT2) by subtracting the
   corresponding proportion of successes (PHAT1, PHAT2)
   from 1.0.

3. Calculate the half-interval (DISP) to the left and right of the
   estimated difference between the proportions, as indicated
   in the given formula.

4. Calculate the left (LOSIDE) side of the confidence interval by subtracting the half-interval (DISP) from the difference between the two proportions. Calculate the right (HISIDE) side of the confidence interval by adding the half-interval (DISP) to the difference between the two proportions.

C. General Program

```
C
C
C     ***************************************************************
C     * THIS PROGRAM COMPUTES A CONFIDENCE INTERVAL FOR THE DIFFERENCE  *
C     * BETWEEN TWO PROPORTIONS BASED UPON LARGE SAMPLES.              *
C     ***************************************************************
C
      INTEGER X1, X2, ALPHA
      REAL LOSIDE, HISIDE
C
C
C     ***************************************************************
C     * STEP 1.  DETERMINE THE SAMPLE SIZES (N1, N2), THE CORRESPONDING *
C     * NUMBER OF SUCCESSES (X1, X2), THE VALUE OF THE STANDARD NORMAL *
C     * RANDOM VARIABLE (ZALFA2) AND THE CONFIDENCE LEVEL (ALPHA).  IN- *
C     * SURE THAT THE SAMPLE SIZES ARE EQUAL TO OR GREATER THAN 30 AND *
C     * THAT THE NUMBER OF SUCCESSES DO NOT EXCEED THE SAMPLE SIZES.   *
C     ***************************************************************
C
      READ (5, 500) N1, X1, N2, X2, ZALFA2, ALPHA
  500 FORMAT (4I6, F10.3, I6)
      IF (N1 .LT. 30 .OR. N2 .LT. 30) STOP
C
C     ***************************************************************
C     * STEP 2.  CALCULATE THE PROPORTION OF SUCCESSES IN EACH SAMPLE  *
C     * (PHAT1, PHAT2) AND THE PROPORTION OF FAILURES IN EACH SAMPLE   *
C     * (QHAT1, QHAT2).                                               *
C     ***************************************************************
C
      PHAT1 = FLOAT (X1) / N1
      PHAT2 = FLOAT (X2) / N2
      QHAT1 = 1.0 - PHAT1
      QHAT2 = 1.0 - PHAT2
C
C     ***************************************************************
C     * STEP 3.  CALCULATE THE DISPLACEMENT (DISP).                   *
C     ***************************************************************
C
      DISP = ZALFA2 * SQRT(((PHAT1 * QHAT1) / N1) + ((PHAT2 * QHAT2) /
     * N2))
C
C     ***************************************************************
C     * STEP 4.  CALCULATE AND PRINT THE CONFIDENCE INTERVAL.         *
C     ***************************************************************
C
      LOSIDE = (PHAT1 - PHAT2) - DISP
      HISIDE = (PHAT1 - PHAT2) + DISP
      WRITE (6, 600) N1, N2
  600 FORMAT (1X,21HTHE SAMPLE SIZES ARE , 2I6)
      WRITE (6, 601) PHAT1, PHAT2
  601 FORMAT (1X,33HTHE PROPORTIONS OF SUCCESSES ARE , 2F10.3)
      WRITE (6, 602) LOSIDE, HISIDE, ALPHA
  602 FORMAT (1X,46HTHE CONFIDENCE INTERVAL FOR THE DIFFERENCE IS ,
     * F10.3, 7H . . . , F10.3, 6H WITH , I6, 13H % CONFIDENCE)
      STOP
      END
```

D. Example Problems

Example 1   A political survey of a major city revealed that 300 adults of a sample of 500 voters favored a liberal candidate while 150 of 290 of the "youth voters" aged 18–21 years favored the same candidate. Calculate a 95 percent confidence interval for the difference between proportions favoring this candidate.

Input Data

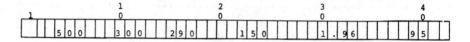

Solution   THE SAMPLE SIZES ARE   500   290
THE PROPORTIONS OF SUCCESSES ARE   0.600   0.517
THE CONFIDENCE INTERVAL FOR THE DIFFERENCE IS 0.011 ...
0.155 WITH 95 PERCENT CONFIDENCE

Example 2   In a study of the winners and losers in two different games at a Las Vegas casino, it showed that 45 out of 110 people won at the dice games, while 30 out of 90 won on the slot machines. What is the 99 percent confidence for the difference between the proportions of winners?

Input Data

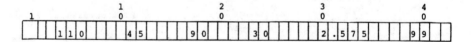

Solution   THE SAMPLE SIZES ARE   110   90
THE PROPORTIONS OF SUCCESSES ARE   0.409   0.333
THE CONFIDENCE INTERVAL FOR THE DIFFERENCE IS −0.100 ...
0.252 WITH 99 PERCENT CONFIDENCE.

A. Statement of Problem

Given a set of N variates $X_1, X_2, \ldots, X_N$, write a computer program which computes a $(1 - \alpha)$ 100 percent confidence interval for the variance σ^2 of an approximately normal distribution as given by the relationship:

$$\frac{(N - 1)S^2}{\chi^2_{(\alpha/2)}} < \sigma^2 < \frac{(N - 1)S^2}{\chi^2_{(1 - \alpha/2)}}$$

where:

$\chi^2_{\alpha/2}$ and $\chi^2_{1 - \alpha/2}$ are the values of a chi square statistic with $N - 1$ degrees of freedom, having areas of $\alpha/2$ and $1 - \alpha/2$, respectively, to the left and to the right of the critical values $(\chi^2_{\alpha/2}$ and $\chi^2_{1 - \alpha/2})$.

B. Algorithm

1. Determine the number (N) of data points (X) in the sample and insure it is 2 or more.

2. Calculate the variance (S²) of the sample by evaluating the formula:

$$S^2 = \frac{N \sum X^2 - (\sum X)^2}{N \cdot (N - 1)}$$

3. Determine the values of the chi square statistic for the left (CHISQL) and right (CHISQR) sides of the confidence interval and the confidence level (ALPHA).

4. Calculate the left side (LOSIDE) of the confidence interval by dividing the sample size minus 1 times the variance (S2) by the value of chi square (CHISQL). Compute the right side (HISIDE) of the confidence interval by dividing the sample size minus 1 times the variance (S2) by the value of chi square (CHISQR).

C. General Program

```
C
C
C     ***********************************************************************
C     * THIS PROGRAM CALCULATES A CONFIDENCE INTERVAL FOR THE VARIANCE    *
C     * OF A POPULATION.                                                   *
C     ***********************************************************************
C
C
      INTEGER ALPHA
      REAL LOSIDE, HISIDE
```

```
C
C
C     ******************************************************************
C     * STEP 1.  DETERMINE THE NUMBER (N) OF DATA POINTS (X) IN THE    *
C     * SAMPLE AND INSURE IT IS TWO OR MORE.                           *
C     ******************************************************************
C
      READ (5, 500) N
 500  FORMAT (I6)
      IF (N .LT. 2) STOP
C
C     ******************************************************************
C     * STEP 2.  CALCULATE THE VARIANCE (S2) OF THE SAMPLE.            *
C     ******************************************************************
C
C
C
C     ******************************************************************
C     * INITIALIZE THE ACCUMULATORS OF THE SUM OF THE DATA POINTS (SUMX)*
C     * AND THE SUM OF THE SQUARED DATA POINTS (SUMXSQ).               *
C     ******************************************************************
C
      SUMX = 0.0
      SUMXSQ = 0.0
C
C     ******************************************************************
C     * READ IN THE DATA POINTS AND ACCUMULATE THEIR SUM (SUMX) AND THE *
C     * SUM OF THEIR SQUARES (SUMXSQ).                                 *
C     ******************************************************************
C
      DO 22 I = 1, N
      READ (5, 501) X
 501  FORMAT (F10.3)
      SUMX = SUMX + X
 22   SUMXSQ = SUMXSQ + (X * X)
C
C     ******************************************************************
C     * CALCULATE THE VARIANCE (S2) OF THE SAMPLE.                     *
C     ******************************************************************
C
      S2 = ((N * SUMXSQ) - (SUMX * SUMX)) / (N * (N - 1))
C
C     ******************************************************************
C     * STEP 3.  DETERMINE THE VALUES OF THE CHI SQUARE STATISTIC FOR  *
C     * THE LEFT (CHISQL) AND THE RIGHT (CHISQR) SIDES OF THE CONFIDENCE*
C     * INTERVAL AND THE CONFIDENCE LEVEL (ALPHA).                     *
C     ******************************************************************
C
      READ (5, 502) CHISQL, CHISQR, ALPHA
 502  FORMAT (2F10.3, I6)
C
C     ******************************************************************
C     * STEP 4.  CALCULATE AND PRINT THE CONFIDENCE INTERVAL.          *
C     ******************************************************************
C
      LOSIDE = ((N - 1) * S2) / CHISQL
      HISIDE = ((N - 1) * S2) / CHISQR
      WRITE (6, 600) S2
 600  FORMAT (1X,30HTHE VARIANCE OF THE SAMPLE IS , F10.3)
      WRITE (6, 601) CHISQL, CHISQR
 601  FORMAT (1X,26HTHE CHI SQUARE VALUES ARE , 2F10.3)
      WRITE (6, 602)
 602  FORMAT (1X,55HTHE CONFIDENCE INTERVAL FOR THE POPULATION VARIANCE
     *IS )
      WRITE (6, 603) LOSIDE, HISIDE, ALPHA
 603  FORMAT (1X, 5X, F10.3, 7H . . . , F10.3, 6H WITH , I6,
     *   13H % CONFIDENCE)
      STOP
      END
```

142

D. Example Problems

The weights of six boxes of cereal are 15.5, 14.8, 15.1, 15.3, 14.9, and 15.0 ounces. Find a 95 percent confidence interval for the actual variance of the weights of all cereal boxes produced by this manufacturer.

Example 1

Input Data

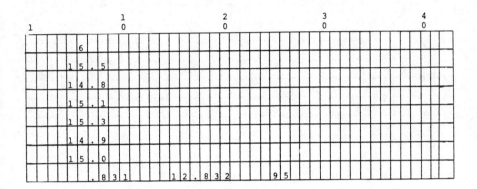

THE VARIANCE OF THE SAMPLES IS 0.068

Solution

THE CHI SQUARE VALUES ARE   0.831   12.832

THE CONFIDENCE INTERVAL FOR THE POPULATION VARIANCE IS 0.410 ... 0.027 WITH 95 PERCENT CONFIDENCE

Ten employees indicated their duration of coffee breaks as: 5, 12, 15, 17, 20, 15, 20, 10, 30, and 18 minutes during a given four-hour period. Find a 99 percent confidence interval for the variance in time of all employees taking coffee breaks.

Example 2

Input Data

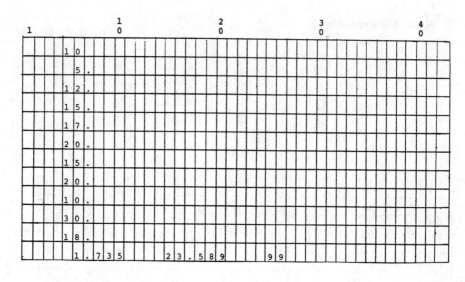

Solution

THE VARIANCE OF THE SAMPLE IS 45.289

THE CHI SQUARE VALUES ARE   1.735   23.589

THE CONFIDENCE INTERVAL FOR THE POPULATION VARIANCE
IS 234.928 ... 17.279 WITH 99 PERCENT CONFIDENCE

1. A sample of 120 variates has a mean of 70 and a standard deviation of 5.5. Find the 98 percent confidence limits for the mean of the population.

2. A random sample of eight cans of peaches from a large shipment showed the following weights (in ounces):

12.1	11.0
10.9	11.9
11.5	12.2
12.3	11.3

   Find a 95 percent confidence interval for the entire shipment of peaches.

3. A trucking company needs to decide whether to purchase brand A or brand B tires. An experiment is conducted using 36 tires of brand A and 48 tires of brand B. The tires are used until they reach $1/8$ in. of treadwear. The results are

   $$\text{Brand A: } \overline{x}_1 = 28{,}750 \text{ miles,} \qquad s_1 = 2000 \text{ miles}$$

   $$\text{Brand b: } \overline{x}_2 = 21{,}540 \text{ miles,} \qquad s_2 = 4000 \text{ miles}$$

   Find the 99 percent confidence interval for the difference between the two population means.

4. Given two samples of size $n_1 = 12$ and $n_2 = 20$ from two independent normal populations with $\overline{x}_1 = 500$, $\overline{x}_2 = 420$, $s_1 = 10$, $s_2 = 20$, find a 95 percent confidence interval for the difference between the two population means.

5. A random sample of 500 families identified a total of 180 families as owners of compact cars. Find a 95 percent confidence interval for the actual proportion of owners of compact cars from the entire city.

6. At a state university, a random sample of 200 men and 200 women were interviewed for drug usage. It was learned that 60 men and 38 women admitted to drug usage at some point in time. Find a 99 percent confidence interval for the difference between all men and women on campus who have indulged in drug usage.

7. Twelve students indicated the time taken for registration at their university was 40, 52, 68, 72, 50, 90, 120, 36, 75, 112, 105, and 48 minutes. Find a 99 percent confidence interval for the time variance in registration time of all students at the university.

Chapter 6

Hypothesis Testing

☆ **Tests Concerning Means (σ known and $N \geq 30$)**

A. Statement of Problem

Suppose we wished to test the hypothesis H_0: $\mu = \mu_0$ against the alternative hypothesis H_1: $\mu \neq \mu_0$ with a given level of significance α. Assume that the distribution of $\bar{x}$ is normally distributed; that we select a random sample of size N from a population with mean μ and standard deviation σ. Using the calculating statistic

$$Z = \frac{\bar{x} - \mu_0}{\sigma/\sqrt{n}}$$

write a computer program which will:

(a) Compute the Z test statistic.

(b) Accept or reject the null hypothesis.

B. Algorithm

1. Determine the sample size (N), the sample mean (XBAR), the sample standard deviation (SIGMA), the confidence level (ALPHA), the corresponding value of the standard normal random variable (ZALFA2), and the hypothetical value of the population mean (MU0). Insure that the sample size is equal to or greater than 30.

2. Compute the test statistic (Z) by evaluating the given formula.

3. Accept the null hypothesis if the absolute value of the test statistic (Z) is less than the value of the standard normal random variable (ZALFA2).

C. General Program

```
C
C
C      ***********************************************************************
C      *  THE NULL HYPOTHESIS TESTED BY THIS PROGRAM IS THAT THE TRUE POP-*
C      *  ULATION MEAN IS EQUAL TO SOME SPECIFIED VALUE.  THE ALTERNATIVE *
C      *  HYPOTHESIS IS THAT THE TRUE POPULATION MEAN IS NOT EQUAL TO THE *
C      *  SPECIFIED VALUE.  THE TEST IS BASED UPON A LARGE SAMPLE WHOSE   *
C      *  MEAN AND STANDARD DEVIATION ARE KNOWN.                          *
C      ***********************************************************************
C
       REAL MU0
C
C      ***********************************************************************
C      *  STEP 1.  DETERMINE THE SAMPLE SIZE (N), THE SAMPLE MEAN (XBAR), *
C      *  THE SAMPLE STANDARD DEVIATION (SIGMA), THE CONFIDENCE LEVEL     *
C      *  (ALPHA), THE CORRESPONDING VALUE OF THE STANDARD NORMAL RANDOM  *
C      *  VARIABLE (ZALFA2) AND THE HYPOTHETICAL POPULATION MEAN (MU0).   *
C      *  INSURE THAT THE SAMPLE SIZE IS EQUAL TO OR GREATER THAN 30.     *
C      ***********************************************************************
C
       READ (5, 500) N, XBAR, SIGMA, ALPHA, ZALFA2, MU0
  500  FORMAT (I6, 5F10.3)
       IF (N.LT.30) STOP
       WRITE (6, 600) N
  600  FORMAT (1X, 19HTHE SAMPLE SIZE IS , I6)
       WRITE (6, 601) XBAR
  601  FORMAT (1X, 19HTHE SAMPLE MEAN IS , F10.3)
       WRITE (6, 602) SIGMA
  602  FORMAT (1X, 33HTHE SAMPLE STANDARD DEVIATION IS , F10.3)
       WRITE (6, 603) MU0
  603  FORMAT (1X, 36HTHE HYPOTHETICAL POPULATION MEAN IS , F10.3)
C
C      ***********************************************************************
C      *  STEP 2.  COMPUTE THE TEST STATISTIC (Z).                        *
C      ***********************************************************************
C
       Z = (XBAR - MU0) / (SIGMA / (N ** .5))
       WRITE (6, 604) Z
  604  FORMAT (1X, 22HTHE TEST STATISTIC IS , F10.3)
C
C      ***********************************************************************
C      *  STEP 3.  ACCEPT OR REJECT THE NULL HYPOTHESIS.                  *
```

```
C     ****************************************************************************
C
      IF (ABS(Z).LT.ZALFA2) GO TO 30
      WRITE (6, 605)
  605 FORMAT (1X, 26HREJECT THE NULL HYPOTHESIS)
      GO TO 40
   30 WRITE (6, 606)
  606 FORMAT (1X, 26HACCEPT THE NULL HYPOTHESIS)
   40 WRITE (6, 607) ALPHA
  607 FORMAT (1X, 12H     AT THE , F10.3, 21HLEVEL OF SIGNIFICANCE)
      STOP
      END
```

D. Example Problems

Example 1   A manufacturer claims the strength of its concrete is 3500 psi (pound per square inch). If a sample of 40 in-place concrete sections yields a mean strength of 3250 psi and a standard deviation of 200, does this support or refute the manufacturer's claim at an $\alpha = 0.05$ level of significance?

Input Data

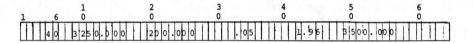

Solution   THE SAMPLE SIZE IS 40

THE SAMPLE MEAN IS 3250.000

THE SAMPLE STANDARD DEVIATION IS 200.000

THE HYPOTHETICAL POPULATION MEAN IS 3500.000

THE TEST STATISTIC IS -7.906

REJECT THE NULL HYPOTHESIS
    AT THE 0.050 LEVEL OF SIGNIFICANCE

Example 2   A manufacturer of golf balls claims that their compression rate is 100 units. A random sample of 50 balls showed the average rate of compression to be 105 units with a standard deviation of 5 units. Can the manufacturer's claim be accepted at the $\alpha = 0.05$ level of significance?

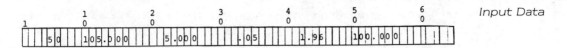

Input Data

THE SAMPLE SIZE IS 50   *Solution*
THE SAMPLE MEAN IS 105.000
THE SAMPLE DEVIATION IS 5.000
THE HYPOTHETICAL MEAN IS 100.000
THE TEST STATISTIC IS 7.071
REJECT THE NULL HYPOTHESIS
       AT THE 0.050 LEVEL OF SIGNIFICANCE

☆ Tests Concerning Means (σ unknown and $N < 30$)

A. Statement of Problem

Suppose we wished to test the hypothesis H_0: $\mu = \mu_0$ against one of the following alternative hypotheses: H: $\mu \neq \mu_0$, H_1': $\mu < \mu_0$, H_1'': $\mu > \mu_0$, at a given level of significance α. Assume that the distribution of $\bar{x}$ is normally distributed and that random samples of size $N < 30$ are taken from a population with mean μ and unknown standard deviation σ. Using the calculating statistic

$$t = \frac{\mu - \mu_0}{s/\sqrt{n}}$$

write a computer program which will:

   (a) Compute the test statistic t.

   (b) Accept or reject the null hypothesis.

B. Algorithm

   1. Determine the sample size (N) and insure it is greater than 1.

   2. Calculate the sample mean (XBAR) and standard deviation (SIGMA). The sample standard deviation is computed by evaluating the formula:

$$s = \sqrt{\frac{N\sum X^2 - (\sum X)^2}{N(N-1)}}$$

3. Determine the program option (OPTION). This option determines the alternative hypothesis against which the null hypothesis is to be treated. A value of 1 indicates that the alternative hypothesis is that the population mean is not equal to some specified value. A value of 2 indicates that the alternative hypothesis is that the population mean is less than some specified value. A value of 3 indicates that the alternative hypothesis is that the population mean is greater than some specified value.

4. Determine the confidence level (ALPHA), the corresponding value of the *T* distribution (T), and the hypothetical population mean (MU∅).

5. Calculate the test statistic (TTEST) by evaluating the given formula.

6. Accept or reject the null hypothesis.

C. General Program

```
C
C     ***************************************************************
C     * THIS PROGRAM TESTS THE HYPOTHESIS THAT THE POPULATION MEAN IS   *
C     * EQUAL TO SOME HYPOTHETICAL VALUE AGAINST ONE OF THE ALTERNATIVE *
C     * HYPOTHESIS 1) THAT THE POPULATION MEAN IS NOT EQUAL TO THE HYPO-*
C     * THETICAL VALUE (H1), 2) THAT THE POPULATION MEAN IS LESS THAN   *
C     * THE HYPOTHETICAL VALUE (H1P), OR 3) THAT THE POPULATION MEAN IS *
C     * GREATER THAN THE HYPOTHETICAL VALUE (H1PP).  THE TEST IS BASED  *
C     * UPON A SMALL SAMPLE.                                           *
C     ***************************************************************
C
      INTEGER ALPHA, OPTION
      REAL MU0
C
C     ***************************************************************
C     * STEP 1.  DETERMINE THE SAMPLE SIZE (N) AND INSURE IT IS GREATER *
C     * THAN 1.                                                        *
C     ***************************************************************
C
      READ (5, 500) N
 500  FORMAT (I6)
      IF (N .LT. 2) STOP
C
C     ***************************************************************
C     * STEP 2.  CALCULATE THE SAMPLE MEAN (XBAR) AND STANDARD DEVIATION*
C     * (SIGMA).                                                       *
C     ***************************************************************
C
C
C
C     ***************************************************************
C     * INITIALIZE THE VARIOUS ACCUMULATORS.                           *
C     ***************************************************************
C
      SUMX = 0.0
      SUMX2 = 0.0
C
C
C     ***************************************************************
C     * SET UP A LOOP TO READ IN ALL N DATA POINTS (X).               *
```

```
C
C      ***********************************************************************
       DO 10 J = 1, N
       READ (5, 501) X
 501   FORMAT (F10.3)
C
C      ***********************************************************************
C      * ACCUMULATE THE SUM OF THE DATA POINTS (SUMX) AND THE SUM OF THE  *
C      * SQUARES OF THE DATA POINTS (SUMX2).                              *
C      ***********************************************************************
C
       SUMX = SUMX + X
 10    SUMX2 = SUMX2 + (X * X)
C
C      ***********************************************************************
C      * COMPUTE THE SAMPLE MEAN (XBAR) AND STANDARD DEVIATION (SIGMA).   *
C      ***********************************************************************
C
       XBAR = SUMX / N
       SIGMA = SQRT(((N * SUMX2) - (SUMX * SUMX)) / (N * (N - 1.0)))
       WRITE (6, 600) XBAR
 600   FORMAT (1X,19HTHE SAMPLE MEAN IS , F10.3)
       WRITE (6, 601) SIGMA
 601   FORMAT (1X,33HTHE SAMPLE STANDARD DEVIATION IS , F10.3)
C
C      ***********************************************************************
C      * STEP 3.  DETERMINE THE PROGRAM OPTION (OPTION).  AN OPTION OF 1 *
C      * INDICATES THAT THE ALTERNATIVE HYPOTHESIS IS H1.  AN OPTION OF 2*
C      * INDICATES THAT THE ALTERNATIVE HYPOTHESIS IS H1P.  A VALUE OF 3 *
C      * INDICATES THAT THE ALTERNATIVE HYPOTHESIS IS H1PP.              *
C      ***********************************************************************
C
       READ (5, 502) OPTION
 502   FORMAT (I6)
C
C      ***********************************************************************
C      * STOP THE PROGRAM IF THE OPTION IS INVALID.                       *
C      ***********************************************************************
C
       GO TO (20, 20, 20), OPTION
       STOP
 20    WRITE (6, 602) OPTION
 602   FORMAT (1X,22HTHE PROGRAM OPTION IS , I6)
C
C      ***********************************************************************
C      * STEP 4.  DETERMINE THE CONFIDENCE LEVEL (ALPHA), THE CORRESPOND-*
C      * ING VALUE OF THE T DISTRIBUTION (T) AND THE HYPOTHETICAL POPULA-*
C      * TION MEAN (MU0).                                                *
C      ***********************************************************************
C
       READ (5, 503) ALPHA, T, MU0
 503   FORMAT (I6, 2F10.3)
       WRITE (6, 603) MU0
 603   FORMAT (1X,36HTHE HYPOTHETICAL POPULATION MEAN IS , F10.3)
       PRINT, *THE HYPOTHETICAL POPULATION MEAN IS*, MU0
C
C      ***********************************************************************
C      * STEP 5.  CALCULATE THE TEST STATISTIC (TTEST).                  *
C      ***********************************************************************
C
       TTEST = (XBAR - MU0) / (SIGMA / N ** .5)
       WRITE (6, 604) TTEST
 604   FORMAT (1X,22HTHE TEST STATISTIC IS , F10.3)
```

151

```
C
C
C      ************************************************************************
C      * STEP 6.  ACCEPT OR REJECT THE NULL HYPOTHESIS.  BRANCH TO THE       *
C      * CORRECT ROUTINE DEPENDING UPON THE PROGRAM OPTION (OPTION).         *
C      ************************************************************************
C
       GO TO (40, 50, 60), OPTION
C
C      ************************************************************************
C      * THIS IS OPTION 1.                                                   *
C      ************************************************************************
C
  40   IF (ABS(TTEST) .LT. T) GO TO 42
  41   WRITE (6, 605)
 605   FORMAT (1X,26HREJECT THE NULL HYPOTHESIS)
       GO TO 100
  42   WRITE (6, 606)
 606   FORMAT (1X,26HACCEPT THE NULL HYPOTHESIS)
       GO TO 100
C
C      ************************************************************************
C      * THIS IS OPTION 2.                                                   *
C      ************************************************************************
C
  50   IF (TTEST .GT. T) GO TO 42
       GO TO 41
C
C      ************************************************************************
C      * THIS IS OPTION 3.                                                   *
C      ************************************************************************
C
  60   IF (TTEST .LT. T) GO TO 42
       GO TO 41
 100   WRITE (6, 607) ALPHA
 607   FORMAT (1X,10H   AT THE , I6, 19H % CONFIDENCE LEVEL)
       STOP
       END
```

D. Example Problems

Example 1   The Strong-Line Company produces fishing line for lake and ocean fishing. Their model 25WX is a 25-pound test line. A sample of eight lines are tested and the mean breaking strengths were found to be 24, 26, 27, 26, 28, 25, 24, and 23 pounds. Using an $\alpha = 0.01$ level of significance, test the manufacturer's claim that $\mu \neq 25$ pounds.

Input Data

1			1 0		2 0		3 0		4 0	

```
      8
   2 4 .
   2 6 .
   2 7 .
   2 6 .
   2 8 .
   2 5 .
   2 4 .
   2 3 .
     1
   9 9        3 . 4 9 9      2 5 .
```

THE SAMPLE MEAN IS 25.375   *Solution*

THE SAMPLE STANDARD DEVIATION IS 1.685

THE PROGRAM OPTION IS 1

THE HYPOTHETICAL POPULATION MEAN IS 25.000

THE TEST STATISTIC IS 0.629

ACCEPT THE NULL HYPOTHESIS AT THE 99 PERCENT CONFI-
DENCE LEVEL

The Always Rite Machine Company produces washers that are sup-   *Example 2*
posed to have a mean thickness of 0.25 cm. A random sample of 20
washers showed thicknesses of 0.26, 0.24, 0.28, 0.27, 0.24, 0.25,
0.28, 0.29, 0.27, 0.28, 0.26, 0.29, 0.28, 0.27, 0.29, 0.29, 0.28. 0.29,
0.24, and 0.28 cm. On the basis of the sample data, is there reason
to suspect that the data are better than the manufacturer's claim
($\alpha = 0.05$)?

Input Data

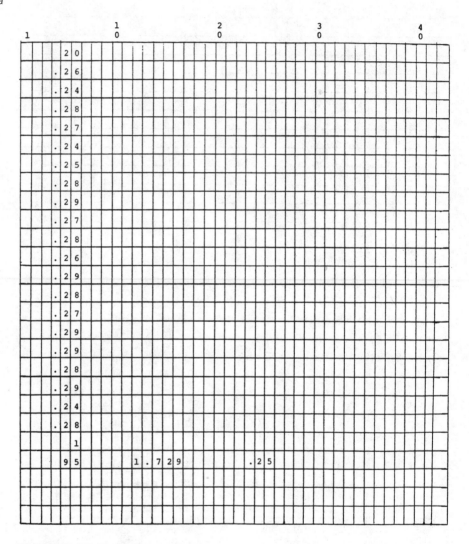

Solution   THE SAMPLE MEAN IS 0.271

THE SAMPLE STANDARD DEVIATION IS 0.018

THE PROGRAM OPTION IS 1

THE HYPOTHETICAL POPULATION MEAN IS 0.250

THE TEST STATISTIC IS 5.478

REJECT THE NULL HYPOTHESIS AT THE 95 PERCENT CONFI-
DENCE LEVEL

☆ Testing the Difference Between Two Means (N_1, $N_2 \geq 30$)

A. Statement of Problem

Suppose we wished to test the null hypothesis H_0: $\mu_1 - \mu_2 = 0$ against the alternative hypothesis H_1: $\mu_1 - \mu_2 \neq 0$. Assume that the distribution of $\bar{x}_1 - \bar{x}_2$ is normally distributed; that random samples of size $N \geq 30$ are taken from two populations with means μ_1 and μ_2 and standard deviations of σ_1 and σ_2. Using the calculating statistic

$$Z = \frac{(\bar{x}_1 - \bar{x}_2) - (\mu_1 - \mu_2)}{\sqrt{\sigma_1^2/N_1 + \sigma_2^2/N_2}} = \frac{(\bar{x}_1 - \bar{x}_2) - d_0}{\sqrt{\sigma_1^2/N_1 + \sigma_2^2/N_2}}$$

where $d_0 = \mu_1 - \mu_2$

write a computer program which will:

(a) Compute the test statistic Z.

(b) Accept or reject the null hypothesis.

B. Algorithm

1. Determine the sample sizes (N1, N2), the sample means (XBAR1, XBAR2), the sample variances (VAR1, VAR2), the confidence level (ALPHA), the corresponding value of the standard normal random variable (ZALFA2), and the specified difference between the two hypothetical means (DZERO). Insure that the sample sizes are equal to or greater than 30.

2. Compute the Z statistic as indicated in the given formula.

3. Accept or reject the null hypothesis. If the absolute value of the computed Z statistic (Z) is less than the value of the standard normal random variable (ZALFA2), the null hypothesis should be accepted. Otherwise, the null hypothesis should be rejected.

C. General Program

```
C
C
C     ***************************************************************
C     * THE NULL HYPOTHESIS TESTED BY THIS PROGRAM IS THAT THE DIFFER- *
C     * ENCE BETWEEN TWO HYPOTHETICAL POPULATION MEANS IS SOME SPECIFIED*
C     * VALUE.   THE ALTERNATIVE HYPOTHESIS IS THAT THE DIFFERENCE IS NOT*
C     * EQUAL TO THE SPECIFIED VALUE.   THE TEST IS BASED UPON TWO LARGE *
C     * SAMPLES WITH KNOWN MEANS AND VARIANCES.                        *
C     ***************************************************************
C
C
C     ***************************************************************
C     * STEP 1.  DETERMINE THE SAMPLE SIZES (N1, N2), THE SAMPLE MEANS  *
C     * (XBAR1, XBAR2) AND VARIANCES (VAR1, VAR2), THE CONFIDENCE LEVEL *
C     * (ALPHA), THE CORRESPONDING VALUE OF THE STANDARD NORMAL RANDOM  *
```

155

```
C     * VARIABLE (ZALFA2) AND THE SPECIFIED DIFFERENCE BETWEEN THE POP- *
C     * ULATION MEANS (DZERO).  INSURE THAT THE SAMPLE SIZES ARE EQUAL   *
C     * TO OR GREATER THAN 30.                                           *
C     *******************************************************************
C
      READ (5, 500) N1, N2, XBAR1, XBAR2, VAR1, VAR2, ALPHA, ZALFA2,
     $     DZERO
500   FORMAT (2I5, 7F10.3)
      IF (N1.LT.30 .OR. N2.LT.30) GO TO 10
      WRITE (6, 600) N1, N2
600   FORMAT (1X, 21HTHE SAMPLE SIZES ARE , I6, 1H , I6)
      WRITE (6, 601) XBAR1, XBAR2
601   FORMAT (1X, 21HTHE SAMPLE MEANS ARE , F10.3, 1H , F10.3)
      WRITE (6, 602) VAR1, VAR2
602   FORMAT (1X, 25HTHE SAMPLE VARIANCES ARE , F10.3, 1H , F10.3)
C
C     *******************************************************************
C     * STEP 2.  COMPUTE THE TEST STATISTIC (Z) AS INDICATED IN THE GIV-*
C     * EN FORMULA.                                                     *
C     *******************************************************************
C
      Z = ((XBAR1 - XBAR2) - DZERO) / SQRT((VAR1 / N1) + (VAR2 / N2))
      WRITE (6, 603) Z
603   FORMAT (1X, 24HTHE TEST STATISTIC Z IS , F10.3)
C
C     *******************************************************************
C     * STEP 3.  ACCEPT OR REJECT THE NULL HYPOTHESIS.                  *
C     *******************************************************************
C
      IF (ABS(Z).LT.ZALFA2) GO TO 10
      WRITE (6,604)
604   FORMAT (1X, 26HREJECT THE NULL HYPOTHESIS)
      GO TO 20
10    WRITE (6, 605)
605   FORMAT (1X, 26HACCEPT THE NULL HYPOTHESIS)
20    WRITE (6, 606) ALPHA
606   FORMAT (1X, 9H   AT THE, F10.3, 22H LEVEL OF SIGNIFICANCE)
      STOP
      END
```

D. Example Problems

Example 1   A national merit scholarship test in history was taken by 1600 students. The boys' average score was 72.7 with a variance of 15.8, while the girls' average score was 75.2 with variance of 27.2. If 600 boys and 1000 girls took the test, determine whether the achievement of the girls was significantly different from that of the boys ($\alpha = 0.05$).

Input Data

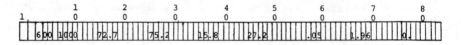

THE SAMPLE SIZES ARE 600   1000   *Solution*

THE SAMPLE MEANS ARE 72.700   75.200

THE SAMPLE VARIANCES ARE 15.800   27.200

THE TEST STATISTIC Z IS -10.805

REJECT THE NULL HYPOTHESIS

    AT THE 0.050 LEVEL OF SIGNIFICANCE

A pen manufacturer claims that his CIB pens write longer than the   *Example 2*
leading competitor, a manufacturer of ball point pens. In a sample of
40 CIB pens, a distance of 2.56 miles was written with a standard
deviation of 0.42 mile, while the leading competitor had corresponding
values of 2.37 and 0.23, respectively. Find whether or not, from a
sample size of 50 tests, significant differences between means exist
if the claims made by CIB are valid ($\alpha = 0.01$).

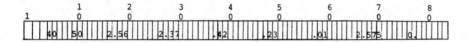

Input Data

THE SAMPLE SIZES ARE 40   50   *Solution*

THE SAMPLE MEANS ARE 2.560   2.370

THE SAMPLE VARIANCES ARE 0.420   0.230

THE TEST STATISTIC Z IS 1.546

ACCEPT THE NULL HYPOTHESIS

    AT THE 0.010 LEVEL OF SIGNIFICANCE

☆ Testing the Difference Between Two Means (N_1, $N_2 < 30$)

A. Statement of Problem

Suppose we wished to test the null hypothesis H_0: $\mu_1 - \mu_2 = d_0$
against one of the following alternative hypotheses H_1: $\mu_1 - \mu_2 \neq$
0, H_1': $\mu_1 - \mu_2 < d_0$, H_1'': $\mu_1 - \mu_2 > d_0$. Assume that the two
populations are approximately normally distributed with equivalent
standard deviations $\sigma_1 = \sigma_2$. Using the calculating statistic

$$t = \frac{(\bar{x}_1 - \bar{x}_2) - d_0}{S_p\sqrt{(1/N_1) + (1/N_2)}}$$

where

$$S_p = \sqrt{\frac{(N_1 - 1)S_1^2 + (N_2 - 1)S_2^2}{N_1 + N_2 - 2}}$$

write a computer program which will:

(a) Compute the test statistic t.

(b) Accept or reject the null hypothesis.

B. Algorithm

1. Determine the first program option (OPT1). A value of 1 will signify that the sample means and variances are known and are given as input. A value of 2 will signify that the input consists of raw data.

2. If the value of OPT1 is 1, read in the sample sizes (N1, N2), the means (XBAR1, XBAR2) and variances (VAR1, VAR2). Then proceed to Step 4.

3. If the value of OPT1 is 2, read in the two samples and calculate their means (XBAR1, XBAR2) and variances (VAR1, VAR2).

4. Calculate the pooled variance (PVAR) by evaluating the given formula.

5. Determine the confidence level (ALPHA), the corresponding tabular value of T (TABT), and the specified difference between means (DZERO).

6. Determine the value of the second program option (OPT2). A value of 1 indicates that the alternative hypothesis is that the difference between means is not equal to the specified value. A value of 2 indicates that the alternative hypothesis is that the difference between means is less than the specified value. A value of 3 indicates that the alternative hypothesis is that the difference between means is greater than the specified value.

7. Calculate the test statistic (T) by evaluating the given formula.

8. Accept or reject the null hypothesis. If the value of OPT2 is 1, the null hypothesis is accepted if the absolute value of T is less than TABT. If the value of OPT2 is 2, the null hypothesis is accepted if the value of T is greater than TABT. If the value of OPT2 is 3, the null hypothesis is accepted if the value of T is less than TABT.

C. General Program

```
C
C      **************************************************************
C      * THE NULL HYPOTHESIS TESTED BY THIS PROGRAM IS THAT THE DIFFER-  *
C      * ENCE BETWEEN TWO POPULATION MEANS IS EQUAL TO SOME SPECIFIED    *
C      * VALUE.  ALTERNATIVE HYPOTHESIS ARE 1) THAT THE DIFFERENCE IS NOT*
C      * EQUAL TO THE SPECIFIED VALUE(H1), 2) THAT THE DIFFERENCE BETWEEN*
C      * MEANS IS LESS THAN THE SPECIFIED VALUE (H1P) AND 3) THAT THE     *
C      * DIFFERENCE BETWEEN MEANS IS GREATER THAN THE SPECIFIED VALUE     *
C      * (H1PP).  THE TEST IS BASED UPON TWO SMALL SAMPLES.              *
C      **************************************************************
C
       INTEGER OPT1, OPT2
C
C      **************************************************************
C      * STEP 1.  DETERMINE THE VALUE OF THE OPT1.  IF THIS VALUE IS 1   *
C      * THE PROGRAM ASSUMES THAT THE SAMPLE MEANS AND VARIANCES ARE      *
C      * KNOWN AND GIVEN AS INPUT.  IF THE VALUE IS 2 THE PROGRAM ASSUMES*
C      * THAT THE INPUT CONSISTS OF RAW DATA.                           *
C      **************************************************************
C
       READ (5, 500) OPT1
  500  FORMAT (I2)
C
C      **************************************************************
C      * IF THE VALUE OF OPT1 IS 1 BRANCH TO THE FIRST OF THE STATEMENT  *
C      * NUMBERS APPEARING INSIDE THE PARENTHESES OF THE FOLLOWING IN-    *
C      * STRUCTION.  IF THE VALUE OF OPT1 IS 2 BRANCH TO THE SECOND       *
C      * STATEMENT NUMBER APPEARING INSIDE THE PARENTHESES.  IF THE VALUE*
C      * OF OPTION IS NEITHER A 1 NOR A 2 THE PROGRAM BRANCHES TO NEITHER*
C      * OF THE GIVEN STATEMENT NUMBERS BUT CONTINUES WITH THE NEXT IN-   *
C      * STRUCTION.                                                      *
C      **************************************************************
C
       GO TO (10, 20), OPT1
       STOP
C
C      **************************************************************
C      * STEP 2.  THIS IS OPTION 1 AND ASSUMES THAT THE SAMPLE SIZES,    *
C      * MEANS AND VARIANCES ARE GIVEN.  READ THESE IN AND THEN BRANCH TO*
C      * STEP 4.                                                        *
C      **************************************************************
C
   10  READ (5, 501) N1, N2, XBAR1, XBAR2, VAR1, VAR2
  501  FORMAT (2I6, 4F10.3)
       GO TO 40
C
C      **************************************************************
C      * STEP 3.  THIS IS OPTION 2.  READ IN THE TWO SAMPLES AND CALCU-  *
C      * LATE THEIR MEANS AND VARIANCES.  LOOP THROUGH STEP 3 TWICE (BE- *
C      * CAUSE THERE ARE TWO SAMPLES).                                  *
C      **************************************************************
C
   20  DO 30 I = 1, 2
C
C      **************************************************************
C      * INITIALIZE THE ACCUMULATORS OF THE SUM OF THE DATA POINTS (SUMX)*
C      * AND THE SUM OF THE SQUARES OF THE DATA POINTS (SUMXSQ).         *
C      **************************************************************
C
```

```
      SUMX = 0.0
      SUMXSQ = 0.0
C
C
C     ************************************************************************
C     * READ IN THE NEXT SAMPLE SIZE.                                       *
C     ************************************************************************
C
      READ (5, 502) N
 502  FORMAT (I6)
C
C
C     ************************************************************************
C     * READ IN THE SAMPLE DATA POINTS AND ACCUMULATE THEIR SUM AND THE     *
C     * SUM OF THEIR SQUARES.                                               *
C     ************************************************************************
C
      DO 15 J = 1, N
      READ (5, 503) X
 503  FORMAT (F10.3)
      SUMX = SUMX + X
 15   SUMXSQ = SUMXSQ + (X * X)
C
C
C     ************************************************************************
C     * CALCULATE THE VARIANCE AND THE MEAN FOR THIS SAMPLE.                *
C     ************************************************************************
C
      VAR = ((N * SUMXSQ) - (SUMX * SUMX)) / (N * (N - 1))
      XBAR = SUMX / N
C
C
C     ************************************************************************
C     * IF THIS IS THE FIRST SAMPLE STORE THE MEAN, VARIANCE AND SAMPLE     *
C     * SIZE IN XBAR1, VAR1 AND N1 RESPECTIVELY.  OTHERWISE, STORE THEM     *
C     * IN XBAR2, VAR2 AND N2 RESPECTIVELY.                                 *
C     ************************************************************************
C
      GO TO (50, 60), I
 50   XBAR1 = XBAR
      VAR1= VAR
      N1 = N
      GO TO 30
 60   XBAR2 = XBAR
      VAR2 = VAR
      N2 = N
 30   CONTINUE
C
C
C     ************************************************************************
C     * STEP 4.  CALCULATE THE POOLED VARIANCE (PVAR).                      *
C     ************************************************************************
C
 40   PVAR =(((N1 - 1) * VAR1) + ((N2 - 1) * VAR2)) / (N1 + N2 - 2)
C
C
C     ************************************************************************
C     * STEP 5.  DETERMINE THE VALUE OF T (TABT) THE CONFIDENCE LEVEL       *
C     * (ALPHA) AND THE SPECIFIED VALUE OF THE DIFFERENCE BETWEEN MEANS     *
C     * (DZERO).                                                            *
C     ************************************************************************
C
      READ (5, 504) TABT, ALPHA, DZERO
 504  FORMAT (3F10.3)
C
C     ************************************************************************
```

```
C      * STEP 6.   DETERMINE THE VALUE OF OPT2.   IF THIS VALUE IS 1 IT    *
C      * SIGNIFIES THAT THE ALTERNATIVE HYPOTHESIS IS THAT THE DIFFERENCE*
C      * BETWEEN MEANS IS NOT EQUAL TO THE SPECIFIED VALUE.  A VALUE OF 2*
C      * SIGNIFIES THAT THE ALTERNATIVE HYPOTHESIS IS THAT THE DIFFERENCE*
C      * BETWEEN MEANS IS LESS THAN THE SPECIFIED VALUE.  A VALUE OF 3   *
C      * SIGNIFIES THAT THE ALTERNATIVE HYPOTHESIS IS THAT THE DIFFERENCE*
C      * BETWEEN MEANS IS GREATER THAN THE SPECIFIED VALUE.  INSURE THAT *
C      * THE VALUE OF OPT2 IS 1, 2, OR 3.                                *
C      ***************************************************************
C
       READ (5, 505) OPT2
 505   FORMAT (I2)
       GO TO (100, 100, 100), OPT2
       STOP
C
C      ***************************************************************
C      * STEP 7.   CALCULATE THE TEST STATISTIC (T).                    *
C      ***************************************************************
C
 100   T = ((XBAR1 - XBAR2) - DZERO) / (SQRT(PVAR) * SQRT((1.0 / N1) +
      $    (1.0 / N2)))
       WRITE (6, 600) N1, N2
 600   FORMAT (1X, 21HTHE SAMPLE SIZES ARE , I6, 1H , I6)
       WRITE (6, 601) XBAR1, XBAR2
 601   FORMAT (1X, 21HTHE SAMPLE MEANS ARE , F10.3, 1H, F10.3)
       WRITE (6, 602) VAR1, VAR2
 602   FORMAT (1X, 25HTHE SAMPLE VARIANCES ARE , F10.3, 1H , F10.3)
       WRITE (6, 603) PVAR
 603   FORMAT (1X, 23HTHE POOLED VARIANCE IS , F10.3)
       WRITE (6, 604) T
 604   FORMAT (1X, 22HTHE TEST STATISTIC IS , F10.3)
C
C      ***************************************************************
C      * STEP 8.   ACCEPT OR REJECT THE NULL HYPOTHESIS.               *
C      ***************************************************************
C
C
C      ***************************************************************
C      * BRANCH DEPENDING UPON THE VALUE OF OPT2.                      *
C      ***************************************************************
C
       GO TO (101, 102, 103), OPT2
C
C      ***************************************************************
C      * THIS IS ALTERNATIVE HYPOTHESIS H1.                            *
C      ***************************************************************
C
 101   IF (ABS(T).LT.TABT) GO TO 65
       WRITE (6, 605)
 605   FORMAT (1X, 26HREJECT THE NULL HYPOTHESIS)
       GO TO 110
 65    WRITE (6, 606)
 606   FORMAT (1X, 26HACCEPT THE NULL HYPOTHESIS)
       GO TO 110
C
C      ***************************************************************
C      * THIS IS ALTERNATIVE HYPOTHESIS H1P.                           *
C      ***************************************************************
C
 102   IF (T.GT.TABT) GO TO 70
```

```
       WRITE (6, 607)
607    FORMAT (1X, 26HREJECT THE NULL HYPOTHESIS)
       GO TO 110
70     WRITE (6, 608)
608    FORMAT (1X, 26HACCEPT THE NULL HYPOTHESIS)
       GO TO 110
C
C
C      ************************************************************
C      * THIS IS ALTERNATIVE HYPOTHESIS H1PP.                    *
C      ************************************************************
C
103    IF (T.LT.TABT) GO TO 80
       WRITE (6, 609)
609    FORMAT (1X, 26HREJECT THE NULL HYPOTHESIS)
       GO TO 110
80     WRITE (6, 610)
610    FORMAT (1X, 26HACCEPT THE NULL HYPOTHESIS)
110    WRITE (6, 611) ALPHA
611    FORMAT (1X, 10H    AT THE , F10.3, 22H LEVEL OF SIGNIFICANCE)
       STOP
       END
```

D. Example Problems

Example 1   A course in statistics was taught to 15 students by a conventional classroom procedure, while a second group of 12 students were given the same course by means of computer-assisted instruction. At the conclusion of the quarter an examination was administered to each group. The group of 15 students obtained an average grade of 83 with a variance of 6, while the group of 12 students scored an average of 87 with a variance of 2. Compute a t statistic and test whether or not the two methods of learning are equal, using a $\alpha = 0.05$ level of significance. (Assume that the two populations are normally distributed with equal variances.)

Input Data

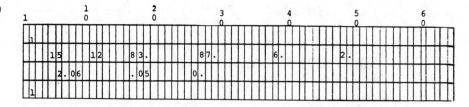

Solution   THE SAMPLE SIZES ARE 15   12

THE SAMPLE MEANS ARE 83.000   87.000

THE SAMPLE VARIANCES ARE 6.000   2.000

THE POOLED VARIANCE IS 4.240
THE TEST STATISTIC IS −5.016
REJECT THE NULL HYPOTHESIS
    AT THE 0.050 LEVEL OF SIGNIFICANCE

Two manufacturers of TV picture tubes were compared for effective-   *Example 2*
ness. A random sample of sets showed the following length of life
as measured in years:

> Brand A: 10,   8,   7,   8,   9,   7,   5,   8,   6,   9
> Brand B: 5,   7,   4,   6,   5,   7,   2,   6,   5,   3

Compute a t statistic and test whether or not the mean lifetime of
manufacturer A exceeds that of manufacturer B ($\alpha = 0.01$). (Assume
that the two populations are normally distributed with equal variances.)

Input Data

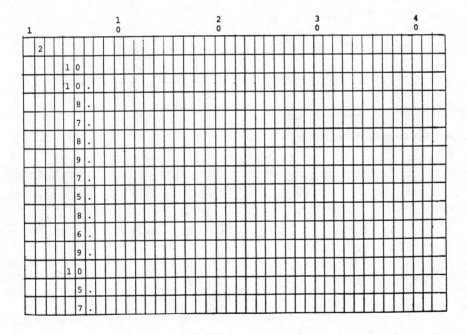

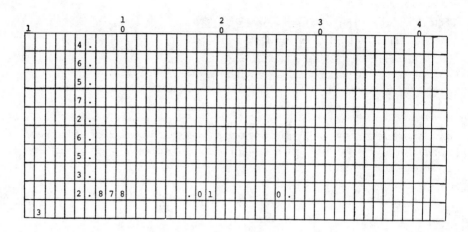

Solution   THE SAMPLE SIZES ARE 10   10
THE SAMPLE MEANS ARE 7.700   5.000
THE SAMPLE VARIANCES ARE 2.233   2.667
THE POOLED VARIANCE IS 2.450
THE TEST STATISTIC IS 3.857
REJECT THE NULL HYPOTHESIS
   AT THE 0.010 LEVEL OF SIGNIFICANCE

☆ **Testing for Independence (Case 1)**

A. Statement of Problem

Suppose we wish to test the hypothesis of whether or not two variables are independent. The null hypothesis assumes that two variables are independent (not related), while the alternative hypothesis assumes that the two variables are dependent (associated). The test for independence uses a χ^2 statistic as given by

$$\chi^2 = \sum_{i=1}^{r} \sum_{j=1}^{c} \frac{(O_{ij} - E_{ij})^2}{E_{ij}}$$

where r is the number of rows and c is the number of columns within an r by c matrix.

   The summation of the differences between the observed (O_{ij}) and expected (E_{ij}) frequencies are to be taken over all cells in the r

by c matrix. The number of degrees of freedom for this test $v = (r - 1)(c - 1)$. Write a computer program which:

(a) Computes the χ^2 statistic.

(b) Accepts or rejects the null hypothesis.

B. Algorithm

1. Determine the number of rows (R) and the number of columns (C) in the contingency table and insure both are greater than 1 but less than or equal to 50. Determine the confidence level (ALPHA) and the corresponding tabular value of the chi square distribution (CHISQT).

2. Read the observed values into the contingency table.

3. Compute and save the sum of each row of observed values, the sum of each column of observed values, and the sum of all observed values.

4. Compute the test statistic (CHISQ). Perform the following steps for all observed values.

   a. Compute the expected value (E_{ij}) by multiplying the total of the row in which this element is positioned by the total of the column in which this element is positioned and then dividing by the total of all observed values.

   b. Accumulate the sum (CHISQ) of the square of the observed value minus the expected value divided by the expected value.

5. If the computed test statistic (CHISQ) is less than the tabular value of the chi square distribution (CHISQT), accept the null hypothesis; otherwise, reject the null hypothesis.

C. General Program

```
C
C
C    *****************************************************************
C    * THIS PROGRAM TESTS FOR THE INDEPENDENCE OF TWO VARIABLES USING A*
C    * A CONTINGENCY TABLE.                                           *
C    *****************************************************************
C
      INTEGER R, C
      DIMENSION O(51, 51)
C
C
C    *****************************************************************
C    * STEP 1.  DETERMINE THE NUMBER OF ROWS (R) AND THE NUMBER OF COL-*
C    * UMNS (C) IN THE CONTINGENCY TABLE.  ALSO DETERMINE THE CONFI-  *
C    * DENCE LEVEL AND THE CORRESPONDING TABULAR VALUE OF THE CHI     *
C    * SQUARE STATISTIC (CHISQT).  INSURE THAT BOTH R AND C ARE GREATER*
C    * THAN 1 AND LESS THAN 51.                                       *
C    *****************************************************************
```

```
C
   1    READ (5, 500) R, C, ALPHA, CHISQT
 500    FORMAT (2I3,  F10.3, F12.5)
        IF (R.LT.2 .OR. R.GT.50 .OR. C.LT.2 .OR. C.GT.50) STOP
C
C       ***********************************************************************
C       * STEP 2.  READ THE OBSERVED VALUES INTO THE CONTINGENCY TABLE       *
C       * (O(I, J)).  THE TABLE IS READ ROW-WISE.                            *
C       ***********************************************************************
C
        READ (5, 501) ((O(I, J), J = 1, C), I = 1, R)
 501    FORMAT (8F10.3)
C
C       ***********************************************************************
C       * STEP 3.  ACCUMULATE THE SUM OF EACH ROW AND THE SUM OF EACH COL-*
C       * UMN.  THE SUM OF THE I TH ROW WILL BE ACCUMULATED IN THE C TH      *
C       * PLUS 1 COLUMN OF THE I TH ROW (O(I, CPLUS1)).  THE SUM OF THE      *
C       * J TH COLUMN WILL BE ACCUMULATED IN THE R TH PLUS 1 ROW OF THE      *
C       * J TH COLUMN (O(RPLUS1, J)).  THE TOTAL OF ALL THE OBSERVATIONS     *
C       * WILL BE ACCUMULATED USING THE R TH PLUS 1, C TH PLUS 1 ELEMENT     *
C       * OF THE CONTINGENCY TABLE (O(RPLUS1, CPLUS1)).                      *
C       ***********************************************************************
C
        CPLUS1 = C + 1
        RPLUS1 = R + 1
C
C       ***********************************************************************
C       * INITIALIZE THOSE ELEMENTS OF THE ARRAY USED TO ACCUMULATE TO-      *
C       * TALS.                                                              *
C       ***********************************************************************
C
        DO 5 I = 1, R
   5    O(I, CPLUS1) = 0.0
        DO 6 J = 1, C
   6    O(RPLUS1, J) = 0.0
        O(RPLUS1, CPLUS1) = 0.0
C
C       ***********************************************************************
C       * SET UP AN OUTER LOOP TO PROCESS EACH ROW.                         *
C       ***********************************************************************
C
        DO 15 I = 1, R
C
C       ***********************************************************************
C       * SET UP AN INNER LOOP TO PROCESS EACH COLUMN.                      *
C       ***********************************************************************
C
        DO 15 J = 1, C
C
C       ***********************************************************************
C       * ACCUMULATE THE ROW TOTALS.                                        *
C       ***********************************************************************
C
        O(I, CPLUS1) = O(I, CPLUS1) + O(I, J)
C
C       ***********************************************************************
C       * ACCUMULATE THE COLUMN TOTALS.                                     *
C       ***********************************************************************
C
        O(RPLUS1, J) = O(RPLUS1, J) + O(I, J)
```

```
C
C      ***************************************************************
C      * ACCUMULATE THE GRAND TOTAL.                                 *
C      ***************************************************************
C
   15   O(RPLUS1, CPLUS1) = O(RPLUS1, CPLUS1) + O(I, J)
C
C      ***************************************************************
C      * STEP 4.  COMPUTE THE TEST STATISTIC (CHISQ).                *
C      * ACCUMULATE THE SUM OF THE OBSERVED VALUES (O(I, J) MINUS THE *
C      * EXPECTED VALUES (EIJ) SQUARED DIVIDED BY THE EXPECTED VALUES *
C      * (EIJ) FOR ALL ELEMENTS OF THE CONTINGENCY TABLE.            *
C      ***************************************************************
C
       WRITE (6, 600)
  600  FORMAT (1X, 34HROW  COLUMN   OBSERVED    EXPECTED)
       CHISQ = 0.0
C
C      ***************************************************************
C      * SET UP AN OUTER LOOP TO PROCESS ALL ROWS.                   *
C      ***************************************************************
C
       DO 20 I = 1, R
C
C      ***************************************************************
C      * SET UP AN INNER LOOP TO PROCESS ALL COLUMNS.                *
C      ***************************************************************
C
       DO 20 J = 1, C
C
C      ***************************************************************
C      * COMPUTE THE EXPECTED VALUE (EIJ) FOR THIS ELEMENT BY MULTIPLYING*
C      * THE ROW TOTAL (O(I, CPLUS1)) BY THE COLUMN TOTAL (O(RPLUS1, C)) *
C      * AND THEN DIVIDING BY THE GRAND TOTAL (O(RPLUS1, CPLUS1)).    *
C      ***************************************************************
C
       EIJ = (O(I, CPLUS1) * O(RPLUS1, J)) / O(RPLUS1, CPLUS1)
       CHISQ = CHISQ + ((O(I, J) - EIJ) ** 2)/ EIJ
   20  WRITE (6, 601) I, J, O(I, J), EIJ
  601  FORMAT (2X, I2, 5X, I2, 2X, F10.3, 2X, F10.3)
       WRITE (6, 602) CHISQ
  602  FORMAT (1X, 36HTHE CHI SQUARE VALUE IS COMPUTED AS , F10.3)
C
C      ***************************************************************
C      * STEP 5.  ACCEPT OR REJECT THE NULL HYPOTHESIS.              *
C      ***************************************************************
C
       IF (CHISQ.LE.CHISQT) GO TO 70
       WRITE (6, 603)
  603  FORMAT (1X, 26HREJECT THE NULL HYPOTHESIS)
       GO TO 90
   70  WRITE (6, 604)
  604  FORMAT (1X, 26HACCEPT THE NULL HYPOTHESIS)
   90  WRITE (6, 605) ALPHA
  605  FORMAT (1X, 10H   AT THE , F10.3, 22H LEVEL OF SIGNIFICANCE)
       STOP
       END
```

D. Example Problems

Example 1   A study of income received from stock by investors and the price of stocks purchased yielded the following table of results:

Income	Price of stock		
	Blue chip	Medium	Low priced
High	972	1720	371
Medium	432	842	647
Low	112	89	317

Test to see if stock income is related to the price of the stock that one invests in (α = 0.05).

Input Data

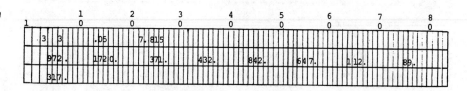

Solution

ROW	COLUMNS	OBSERVED	EXPECTED
1	1	972.000	843.967
1	2	1720.000	1475.829
1	3	371.000	743.203
2	1	432.000	529.305
2	2	842.000	925.585
2	3	647.000	466.109
3	1	112.000	142.728
3	2	89.000	249.585
3	3	317.000	125.687

THE CHI SQUARE VALUE IS COMPUTED AS 743.002
REJECT THE NULL HYPOTHESIS
    AT THE 0.050 LEVEL OF SIGNIFICANCE

Example 2   A study of smoking habits versus income level led to the following table:

	High income	Medium income	Low income
Nonsmokers	72	112	247
Previous smokers	142	82	71
Smokers	219	174	351

Test to see if income level and smoking habits are associated (α = 0.01).

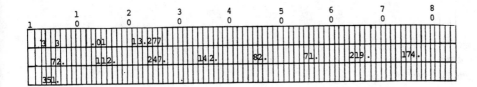

Input Data

Solution

ROW	COLUMN	OBSERVED	EXPECTED
1	1	72.000	126.954
1	2	112.000	107.897
1	3	247.000	196.149
2	1	142.000	86.895
2	2	82.000	73.850
2	3	71.000	134.255
3	1	219.000	219.151
3	2	174.000	186.253
3	3	351.000	338.596

THE CHI SQUARE VALUE IS COMPUTED AS 104.036
REJECT THE NULL HYPOTHESIS
    AT THE 0.010 LEVEL OF SIGNIFICANCE

☆ **Testing for Independence (Case 2)**

A. Statement of Problem

One test of whether two variables are independent uses a 2 by 2 contingency table as given by the following:

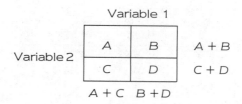

where A, B, C, and D are the observed values; $A + B$, $C + D$, $A + C$, $B + D$ are the marginal totals; $N = A + B + C + D$. Using the identical hypotheses as in Case 1, the χ^2 test statistic is given as

$$\chi^2 = \frac{N(A \cdot D - B \cdot C)^2}{(A + B)(C + D)(A + C)(B + D)}$$

with $\nu = (r - 1)(c - 1)$ degrees of freedom. Write a computer program which:

(a) Computes the χ^2 statistic.

(b) Accepts or rejects the null hypothesis (see algorithm)

B. Algorithm

1. Determine the values of A, B, C, and D.

2. Compute the test statistic (CHISQ) by evaluating the given formula.

3. Determine the confidence level (ALPHA) and the corresponding tabular value of the chi square distribution (CHISQT).

4. Accept or reject the null hypothesis. If the test statistic (CHISQ) is less than the tabular value of the chi square distribution, accept the null hypothesis; otherwise, reject the null hypothesis.

C. General Program

```
C
C
C     ************************************************************************
C     * THIS PROGRAM TESTS THE INDEPENDENCE OF TWO VARIABLES USING A 2 X*
C     * 2 CONTINGENCY TABLE.                                            *
C     ************************************************************************
C
C
C
C     ************************************************************************
C     * STEP 1.  DETERMINE THE VALUES OF A, B, C, AND D.                *
C     ************************************************************************
C
      READ (5, 500) A, B, C, D
  500 FORMAT (4F10.3)
      WRITE (6, 600)
```

```
  600   FORMAT (1X, 24HTHE CONTINGENCY TABLE IS)
        WRITE (6, 601) A, B
  601   FORMAT (1X, F10.3, 3X, F10.3)
        WRITE (6, 601) C, D
C
C       ******************************************************************
C       * STEP 2.  COMPUTE THE TEST STATISTIC (CHISQ).                   *
C       ******************************************************************
C
        CHISQ = ((A + B + C + D) * ((A * D - B * C) ** 2)) / ((A + B ) *
       $   (C + D) * (A + C) * (B + D))
        WRITE (6, 602) CHISQ
  602   FORMAT (1X, 22HTHE TEST STATISTIC IS , F10.3)
C
C       ******************************************************************
C       * STEP 3.  DETERMINE THE CONFIDENCE LEVEL (ALPHA) AND THE CORRE- *
C       * SPONDING TABULAR VALUE OF THE CHI SQUARE DISTRIBUTION (CHISQT). *
C       ******************************************************************
C
        READ (5, 501) ALPHA, CHISQT
  501   FORMAT (F10.3, F12.5)
        WRITE (6, 603) CHISQT
  603   FORMAT (1X, 35HTHE TABULAR VALUE OF CHI SQUARE IS , F10.3)
C
C       ******************************************************************
C       * STEP 4.  ACCEPT OR REJECT THE NULL HYPOTHESIS.                 *
C       ******************************************************************
C
        IF (CHISQ.LT.CHISQT) GO TO 20
        WRITE (6, 604)
  604   FORMAT (1X, 26HREJECT THE NULL HYPOTHESIS)
        GO TO 30
  20    WRITE (6, 605)
  605   FORMAT (1X, 26HACCEPT THE NULL HYPOTHESIS)
  30    WRITE (6, 606) ALPHA
  606   FORMAT (1X, 11H    AT THE , F10.3, 22H LEVEL OF SIGNIFICANCE)
        STOP
        END
```

D. Example Problems

Example 1

The number of men and women who experienced air sickness during an airplane trip where turbulent weather conditions existed were noted as follows:

	Number affected	Number not affected
Men	20	10
Women	15	25

Test to determine whether or not air sickness is related to sex ($\alpha = 0.01$).

Input Data

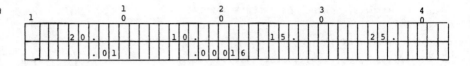

Solution   THE CONTINGENCY TABLE IS
      20.000   10.000
      15.000   25.000
THE TEST STATISTIC IS 5.833
THE TABULAR VALUE OF CHI SQUARE IS 0.000
REJECT THE NULL HYPOTHESIS
    AT THE 0.010 LEVEL OF SIGNIFICANCE

Example 2   A survey among 150 adults identified the following number as being smokers or nonsmokers:

	Male	Female
Smokers	30	80
Nonsmokers	20	20

Is there a relationship between sex and smoking ($\alpha = 0.05$)?

Input Data

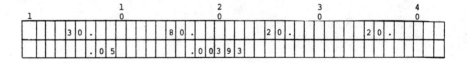

Solution   THE CONTINGENCY TABLE IS
      30.000   80.000
      20.000   20.000
THE TEST STATISTIC IS 6.818
THE TABULAR VALUE OF CHI SQUARE IS 0.004
REJECT THE NULL HYPOTHESIS
    AT THE 0.050 LEVEL OF SIGNIFICANCE

1. A United States government official issued the following statement: "The mean weekly food expenditure for a family of four persons within a large urban city was $75." A sample of 100 such families showed the average amount of food expenditures to be $82. Can we accept the official's statement at the $\alpha = 0.05$ level of significance?

2. A foreign manufacturer of automobiles stated that the company's improved suspension system required only two wheel alignments per year. A sample of seven automobiles showed the following number of wheel alignments per year: 1, 2, 3, 1, 2, 1, and 3. Can we accept the manufacturer's claim at the $\alpha = 0.05$ level of significance?

3. Two types of charge accounts were used by a local department store. Accounts A and B had 50 and 80 charges, average dollar payments of $86.42 and $75.21, and standard deviations of $4.72 and $5.65, respectively. Test at the $\alpha = 0.05$ level of significance whether the average dollar payments of the two populations are equivalent.

4. A retail chain had two kinds of canned fruit, namely brand A and brand B. Consumer preferences during a one-week period indicated the following number of cans sold:

Brand A	50	61	75	48	47	101	112
Brand B	65	82	78	63	55	142	135

   Determine whether there exists a consumer preference for one brand over another ($\alpha = 0.05$).

5. The following table indicated the relationship between job rating and test scores above and below the median:

Test Scores	Job Rating Below average	Average	Above average
Below median	12	8	5
Above median	5	10	20

   Test the hypothesis that job rating is independent of examination results ($\alpha = 0.05$).

6. The following data relate to patients under treatment within a mental hospital:

	Improvement	No improvement
Treatment A	15	35
Treatment B	6	15

Test the hypothesis that recovery is independent of method of treatment.

Chapter 7

Regression and Correlation

☆ **Linear Regression Model**

A. Statement of Problem

Write a program which will compute the following:

1. The least squares estimators a and b for the population parameters α and β as given by:

$$b = \frac{N\sum XY - \sum X \sum Y}{N\sum X^2 - (\sum X)^2}$$

$$a = \bar{Y} - b\bar{X}$$

2. *The standard error of estimate, S, as given by the equation:*

$$S = \sqrt{\frac{N-1}{N-2}(S_Y^2 - b^2 S_X^2)}$$

where

$$S_X^2 = \frac{N \sum X^2 - (\sum X)^2}{N(N-1)}$$

and

$$S_Y^2 = \frac{N \sum Y^2 - (\sum Y)^2}{N(N-1)}$$

3. A confidence interval for the parameter in the linear regression model $\mu_{Y/X} = \alpha + \beta X$ as given by the equation

$$a - \frac{t_{\alpha/2} S \sqrt{\sum X_i^2}}{S_X \sqrt{N(N-1)}} < \alpha < a + \frac{t_{\alpha/2} S \sqrt{\sum X_i^2}}{S_X \sqrt{N(N-1)}}$$

where $t_{\alpha/2}$ is the value of the t statistic with $\nu = N - 2$ degrees of freedom, $\sum X_i^2$ is the sum of the squares of the X variates, S is the unbiased estimator of σ and is given as:

$$S^2 = \frac{N-1}{N-2}(S_Y^2 - b S_X^2)$$

where N is the number of paired variates in the sample.

4. A confidence interval for the parameter in the linear regression model $\mu_{Y/X} = \alpha + \beta X$ as given by the equation

$$b - \frac{t_{\alpha/2} \cdot S}{S_X \sqrt{(N-1)}} < \beta < b + \frac{t_{\alpha/2} \cdot S}{S_X \sqrt{(N-1)}}$$

where b is the point estimator of β and $t_{\alpha/2}$ is the value of the t statistic with $\nu = N - 2$ degrees of freedom, S_X is the standard deviation of the X variates.

B. Algorithm

1. Determine the number (N) of pairs of data points (Y, X); insure that the number of pairs is greater than 1.
2. Accumulate the sum of the values of X (SUMX), the sum of the values of Y (SUMY), the sum of the squares of the values of X (SUMX2), the sum of the squares of the values of Y (SUMY2), and the sum of the cross products of the X and Y values (SUMXY).

3. Calculate the linear regression parameters a and b (A, B) by evaluating the given formula.

4. Calculate the standard error of estimate (S) by evaluating the given formula.

5. Calculate the confidence interval for the parameter a.

   a. Determine the desired value of $t_{\alpha/2}$ (TALFA2) and the corresponding confidence level (ALPHA).

   b. Evaluate the given formula for the confidence interval for the parameter a.

6. Calculate the confidence interval for the parameter b by evaluating the given formula.

C. General Program

```
C
C     ***************************************************************
C     * THIS PROGRAM COMPUTES POINT AND INTERVAL ESTIMATES FOR THE LIN- *
C     * EAR REGRESSION PARAMETERS A AND B AS WELL AS THE STANDARD ERROR *
C     * OF ESTIMATE.                                                 *
C     ***************************************************************
C
      INTEGER ALPHA
      REAL LOSIDE
C
C     ***************************************************************
C     * STEP 1.  DETERMINE THE NUMBER (N) OF Y AND X PAIRS AND INSURE IT*
C     * IS GREATER THAN 1.                                           *
C     ***************************************************************
C
      READ (5, 500) N
  500 FORMAT (I6)
      IF (N .LT. 2) STOP
C
C     ***************************************************************
C     * STEP 2.  ACCUMULATE ALL OF THE REQUIRED SUMS.               *
C     ***************************************************************
C
C
C     ***************************************************************
C     * INITIALIZE THE VARIOUS ACCUMULATORS BY SETTING THEM EQUAL TO *
C     * ZERO.                                                       *
C     ***************************************************************
C
      SUMX = 0.0
      SUMY = 0.0
      SUMX2 = 0.0
      SUMY2 = 0.0
      SUMXY = 0.0
C
C     ***************************************************************
C     * READ IN ALL Y AND X PAIRS AND ACCUMULATE THE VARIOUS SUMS.  *
C     ***************************************************************
C
      DO 10 I = 1, N
      READ (5, 501) Y, X
```

```
  501   FORMAT (2F10.3)
        SUMX = SUMX + X
        SUMY = SUMY + Y
        SUMX2 = SUMX2 + (X * X)
        SUMY2 = SUMY2 + (Y * Y)
   10   SUMXY = SUMXY + (X * Y)
C
C
C     *****************************************************************
C     * STEP 3.  COMPUTE THE POINT ESTIMATES OF THE PARAMETERS A AND B. *
C     *****************************************************************
C
        B = ((N * SUMXY) - (SUMX * SUMY)) / ((N * SUMX2) - (SUMX * SUMX))
        A = (SUMY / N) - (B * (SUMX / N))
        WRITE (6, 600) A
  600   FORMAT (1X,18HTHE PARAMETER A = , F10.3)
C
C     *****************************************************************
C     * STEP 4.  CALCULATE THE STANDARD ERROR OF ESTIMATE (S).       *
C     *****************************************************************
C
C
C
C     *****************************************************************
C     * COMPUTE THE VARIANCE OF THE X VALUES (SX2) AND OF THE Y VALUES *
C     * (SY2).                                                        *
C     *****************************************************************
C
        SX2 = ((N * SUMX2) - (SUMX * SUMX)) / (N * (N - 1))
        SY2 = ((N * SUMY2) - (SUMY * SUMY)) / (N * (N - 1))
C
C     *****************************************************************
C     * COMPUTE THE STANDARD ERROR OF ESTIMATE (S).                  *
C     *****************************************************************
C
        S = SQRT(((N - 1.0) / (N - 2)) * (SY2 - ((B * B) * SX2)))
C
C     *****************************************************************
C     * STEP 5.  CALCULATE THE CONFIDENCE INTERVAL FOR THE PARAMETER A. *
C     *****************************************************************
C
C
C
C     *****************************************************************
C     * STEP 5(A).  READ IN THE CONFIDENCE LEVEL (ALPHA) AND THE CORRES-*
C     * PONDING VALUE OF T (TALFA2).                                 *
C     *****************************************************************
C
        READ (5, 502) TALFA2, ALPHA
  502   FORMAT (F10.3, I6)
C
C     *****************************************************************
C     * STEP 5(B).  EVALUATE THE FORMULA FOR THE CONFIDENCE INTERVAL FOR*
C     * A.                                                           *
C     *****************************************************************
C
        DISP = (TALFA2 * S * SQRT(SUMX2)) / (SQRT(SX2) * SQRT(N * (N -
      *   1.0)))
        LOSIDE = A - DISP
        HISIDE = A + DISP
        WRITE (6, 601) LOSIDE, HISIDE, ALPHA
  601   FORMAT (1X,29HCONFIDENCE INTERVAL FOR A IS , F10.3, 7H . . . ,
      *   F10.3, 6H WITH , I6, 13H % CONFIDENCE)
C
C     *****************************************************************
C     * STEP 6.  EVALUATE THE FORMULA FOR THE CONFIDENCE INTERVAL FOR B.*
C     *****************************************************************
```

178

```
C
      DISP = (TALFA2 * S) / (SQRT(SX2) * SQRT(N - 1.0))
      LOSIDE = B - DISP
      HISIDE = B + DISP
      WRITE (6, 602) B
 602  FORMAT (1X,18HTHE PARAMETER B = , F10.3)
      WRITE (6, 603) LOSIDE, HISIDE, ALPHA
 603  FORMAT (1X,29HCONFIDENCE INTERVAL FOR B IS , F10.3, 7H . . . ,
     *    F10.3, 6H WITH , I6, 13H % CONFIDENCE)
      WRITE (6, 604) S
 604  FORMAT (1X,38HTHE STANDARD ERROR OF THE ESTIMATE IS , F10.3)
      STOP
      END
```

D. Example Problems

Example 1

Given the following data:

X	Y
5	1
6	3
7	1
8	4
9	2
10	3
11	5

find:

(a)   The point estimators a and b of the linear regression model
Y = a + bX.

(b)   The standard error of estimate.

(c)   The 95 percent confidence intervals for α and β.

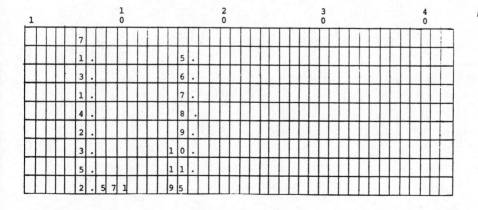

Input Data

Solution   THE PARAMETER A = −1.000

CONFIDENCE INTERVAL FOR A IS −5.872 ... 3.872 WITH 95 PER-
CENT CONFIDENCE

THE PARAMETER B = 0.464

CONFIDENCE INTERVAL FOR B IS −0.127 ... 1.055 WITH 95 PER-
CENT CONFIDENCE

THE STANDARD ERROR OF THE ESTIMATE IS 1.216

Example 2   Seven families listed their annual income X in thousands of dollars
and yearly food expenditures Y in thousands of dollars as follows:

X	Y
14.	2.
8.	1.
4.	1.
11.	1.
26.	3.
17.	2.
18.	2.

Find:

(a)   The equation of the linear regression model $Y = a + bX$ by
solving for the point estimators a and b.

(b)   The standard error of estimate.

(c)   The 99 percent confidence intervals for α and β.

Input Data

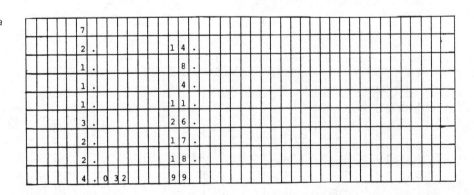

THE PARAMETER A = 0.332   *Solution*

CONFIDENCE INTERVAL FOR A IS −0.626 ... 1.290 WITH 99 PER-
CENT CONFIDENCE

THE PARAMETER B = 0.099

CONFIDENCE INTERVAL FOR B IS 0.037 ... 0.160 WITH 99 PERCENT
CONFIDENCE

THE STANDARD ERROR OF THE ESTIMATE IS 0.271

☆ Exponential Regression Model

A. Statement of Problem

Write a computer program which will calculate the point estimators
c and d of the exponential model, $Y = cd^X$. Taking the logarithm of
both sides of this equation, we have

$$\log Y = \log cd^X$$

Rewriting this equation, we have

$$\log Y = \log c + X \log d$$

Notice the similarity between this equation and the form of the
general linear model as given by

$$Y = a + bX$$

Suppose we let $A = \log c$, $B = \log d$, and $Y = \log y$. Substituting into
the general equations for our linear estimators, we have

$$b = \frac{N \sum X \log Y - \sum X \sum \log Y}{N \sum X^2 - (\sum X)^2}$$

$$a = \overline{\log Y} - b\overline{X}$$

Once a and b are determined, then the equations

$$a = \log c$$

$$b = \log d$$

need to be solved for c and d, respectively.

B. Algorithm

1. Determine the number (N) of pairs of Y and X values; insure
    that this number is greater than 1.

2. Accumulate the sum of the X values (SUMX), the sum of the squares of the X values (SUMX2), the sum of the logarithms of the Y values (SUMLGY), and the sum of the cross product of the X values and the logarithms of the Y values (SMXLGY).

3. Estimate the parameters a and b by evaluating the given formulas.

4. Calculate the estimate of the exponential regression parameters c and d by taking the antilogarithms of a and b, respectively.

C. General Program

```
C
C
C     ***************************************************************
C     * THIS PROGRAM ESTIMATES THE EXPONENTIAL REGRESSION PARAMETERS C  *
C     * AND D.                                                       *
C     ***************************************************************
C
C
C
C     ***************************************************************
C     * STEP 1.  DETERMINE THE NUMBER (N) OF PAIRS Y AND X VALUES AND *
C     * INSURE IT IS GREATER THAN 1.                                 *
C     ***************************************************************
C
      READ (5, 500) N
  500 FORMAT (I6)
      IF (N .LT. 2) STOP
C
C     ***************************************************************
C     * STEP 2.  ACCUMULATE THE REQUIRED SUMS.                       *
C     ***************************************************************
C
C
C
C     ***************************************************************
C     * INITIALIZE THE VARIOUS ACCUMULATORS.                         *
C     ***************************************************************
C
      SUMX = 0.0
      SUMX2 = 0.0
      SUMLGY = 0.0
      SMXLGY = 0.0
C
C     ***************************************************************
C     * READ IN ALL PAIRS OF DATA (Y, X) AND ACCUMULATE THE REQUIRED *
C     * SUMS.                                                        *
C     ***************************************************************
C
      DO 25 K = 1, N
      READ (5, 501) Y, X
  501 FORMAT (2F10.3)
      SUMX = SUMX + X
      SUMX2 = SUMX2 + (X * X)
C
C     ***************************************************************
C     * THE FUNCTION ALOG10 IS USED TO OBTAIN THE LOGARITHM OF Y TO THE *
C     * BASE 10.                                                     *
C     ***************************************************************
C
```

```
C
      SUMLGY = SUMLGY + ALOG10(Y)
  25  SMXLGY = SMXLGY + (X * ALOG10(Y))
C
C
C     ***************************************************************
C     * STEP 3.  ESTIMATE THE PARAMETERS A AND B BY EVALUATING THE GIVEN*
C     * FORMULAS.                                                    *
C     ***************************************************************
C
      B = ((N * SMXLGY) - (SUMX * SUMLGY)) / ((N * SUMX2) - (SUMX *
     *   SUMX))
      A = (SUMLGY / N) - (B * (SUMX / N))
C
C     ***************************************************************
C     * STEP 4.  CALCULATE THE PARAMETERS C AND D.                  *
C     ***************************************************************
C
C
C     ***************************************************************
C     * DERIVE THE ANTILOGARITHM OF A BY RAISING 10 TO THE A-TH POWER. *
C     * THIS GIVES THE ESTIMATE OF THE PARAMETER C.                 *
C     ***************************************************************
C
      C = 10 ** A
C
C     ***************************************************************
C     * DERIVE THE ANTILOGARITHM OF B GIVING THE ESTIMATE OF THE PARA- *
C     * METER D.                                                    *
C     ***************************************************************
C
      D = 10 ** B
      WRITE (6, 600) C
  600 FORMAT (1X,42HTHE EXPONENTIAL REGRESSION PARAMETER C IS , F10.3)
      WRITE (6, 601) D
  601 FORMAT (1X,42HTHE EXPONENTIAL REGRESSION PARAMETER D IS , F10.3)
      STOP
      END
```

D. Example Problems

The enrollment at a particular college in the midwestern portion of   *Example 1*
the United States during the past five years is indicated below:

X (Years)	Y (Enrollment)
1	280
2	360
3	452
4	571
5	725

Use the method of least squares to fit a curve of the form $\mu_{Y/X} = cd^X$.

Input Data

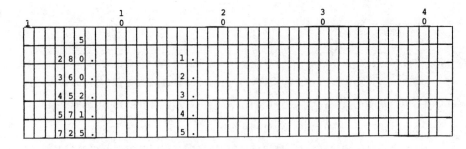

Solution   THE EXPONENTIAL REGRESSION PARAMETER C IS 222.387
THE EXPONENTIAL REGRESSION PARAMETER D IS 1.267

Example 2   The total dollar sales (in thousand-dollar units) in a five-year period for nursery stock is listed below:

X Year	Y Total dollar sales
1	11
2	17
3	29
4	48
5	72

Use the method of least squares to fit a curve of the form $\mu_{Y/X} = cd^X$.

Input Data

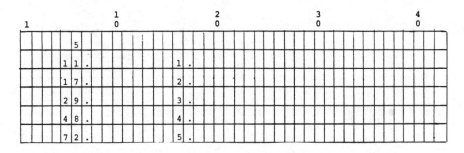

Solution   THE EXPONENTIAL REGRESSION PARAMETER C IS 6.757
THE EXPONENTIAL REGRESSION PARAMETER D IS 1.615

A. Statement of Problem

Given a set of random variables X and Y, develop a computer program which will measure the linear association between the two variables X and Y. This value is referred to as the correlation coefficient and is symbolized by the letter r. This measure of linear association can be evaluated by the following formula:

$$r = \frac{N\sum XY - \sum X\sum Y}{\sqrt{(N\sum X^2 - (\sum X)^2) \cdot (N\sum Y^2 - (\sum Y)^2)}}$$

where N refers to the number of paired variables of X and Y.

B. Algorithm

1. Determine the number of pairs (N) of data points (X, Y) to be used in the analysis. Insure that the number of pairs (N) is two or more.
2. Do the following for all pairs of data points (X, Y).
   a. Accumulate the sum of the X element (SUMX).
   b. Accumulate the sum of the squares of the X element (SUMX2).
   c. Accumulate the sum of the Y element (SUMY).
   d. Accumulate the sum of the squares of the Y element (SUMY2).
   e. Accumulate the sum of the product of X and Y elements (SUMXY).
3. Calculate the numerator (RNUM) of the formula.
4. Calculate the denominator (RDENOM) of the formula.
5. Calculate Pearson's product moment correlation coefficient (R) by dividing the numerator (RNUM), derived in Step 3, by the denominator (RDENOM) derived in Step 4.

C. General Program

```
C
C
C     **************************************************************
C     * THIS PROGRAM CALCULATES THE PEARSON PRODUCT MOMENT CORRELATION *
C     * COEFFICIENT.                                                *
C     **************************************************************
C
C
C     **************************************************************
C     * STEP 1.  DETERMINE THE NUMBER (N) OF PAIRS OF VALUES (X, Y) AND *
C     * INSURE THAT THIS NUMBER IS 2 OR MORE.                       *
```

185

```
C     ******************************************************************
C
      READ (5, 500) N
 500  FORMAT (I6)
      IF (N .LT. 2) STOP
C
C     ******************************************************************
C     * STEP 2.  ACCUMULATE THE VARIOUS SUMS INDICATED IN THE ALGORITHM.*
C     ******************************************************************
C
C
C     ******************************************************************
C     * INITIALIZE THE VARIOUS ACCUMULATORS USED IN THE COMPUTATIONS.  *
C     ******************************************************************
C
      SUMX = 0.0
      SUMY = 0.0
      SUMX2 = 0.0
      SUMY2 = 0.0
      SUMXY = 0.0
C
C     ******************************************************************
C     * REPETITIVELY CYCLE THROUGH THE SET OF INSTRUCTIONS FROM THE ONE *
C     * FOLLOWING THE DO STATEMENT THROUGH AND INCLUDING THE STATEMENT  *
C     * NUMBERED 10.  THIS SET OF INSTRUCTIONS WILL BE CYCLED THROUGH N *
C     * TIMES.                                                          *
C     ******************************************************************
C
      DO 10 I = 1, N
C
C     ******************************************************************
C     * GET THE NEXT PAIR OF VALUES (X, Y).                            *
C     ******************************************************************
C
      READ (5, 501) X, Y
 501  FORMAT (2F10.3)
C
C     ******************************************************************
C     * ACCUMULATE THE VARIOUS SUMS.                                   *
C     ******************************************************************
C
      SUMX = SUMX + X
      SUMY = SUMY + Y
      SUMX2 = SUMX2 + (X * X)
      SUMY2 = SUMY2 + (Y * Y)
      SUMXY = SUMXY + (X * Y)
 10   CONTINUE
C
C     ******************************************************************
C     * STEP 3.  CALCULATE THE NUMERATOR (RNUM) OF THE FORMULA.        *
C     ******************************************************************
C
      RNUM = (N * SUMXY) - (SUMX * SUMY)
C
C     ******************************************************************
C     * STEP 4.  CALCULATE THE DENOMINATOR (RDENOM) OF THE FORMULA.    *
C     ******************************************************************
C
      RDENOM = SQRT (((N * SUMX2 - (SUMX ** 2.0)) * (N * SUMY2 - (SUMY
     *   ** 2.0))))
C
C     ******************************************************************
C     * STEP 5.  CALCULATE THE PEARSON PRODUCT MOMENT CORRELATION      *
C     * COEFFICIENT (R).                                               *
```

```
C    ************************************************************
C
     R = RNUM / RDENOM
     WRITE (6, 600) R
600  FORMAT (1X,54HTHE PEARSON PRODUCT MOMENT CORRELATION COEFFICIENT I
    *S , F10.3)
     STOP
     END
```

D. Example Problems

Example 1

The number of hours studied per week by a group of psychology students along with their accompanying final examination grades are recorded below:

Hours studied/week	Final grade
10	85
5	70
13	76
20	81
14	68
9	95

Find the Pearson's product moment correlation coefficient for the above data.

Input Data

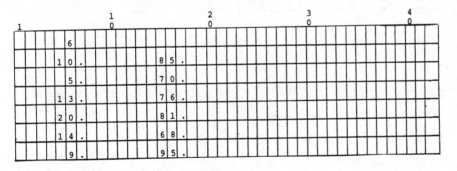

Solution

THE PEARSON PRODUCT MOMENT CORRELATION COEFFICIENT IS −0.023

Example 2

The number of defective parts and the number of total parts produced are listed for seven production workers as follows:

Number of Defective Parts	Total Parts Produced
36	118
25	111
12	136
17	162
24	98
29	154
9	101

Find the Pearson's product moment correlation coefficient for the above data.

Input Data

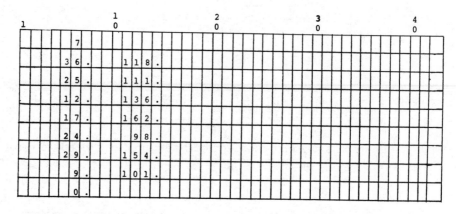

Solution   THE PEARSON PRODUCT MOMENT CORRELATION COEFFICIENT IS 0.019

☆ Spearman's Rank Correlation Coefficient

A. Statement of Problem

Given a set of N ranked values $X_1, X_2, \ldots, X_N$ and a set of ranked values $Y_1, Y_2, \ldots, Y_N$, develop a computer program which will calculate Spearman's rank correlation coefficient, ρ, as given by

$$\rho = 1 - \frac{6 \sum_{i=1}^{N} d_i^2}{N(N^2 - 1)}$$

where $\sum d_i^2$ is the sum of the squared differences over a set of N pairs of observations.

B. Algorithm

1. Determine the number of paired ranks (N) to be used in the test. Insure that the number of paired ranks (N) is 2 or more.

2. Accumulate the sum of the squared differences in rank (SSQDIF) for all N paired ranks (X, Y).

3. Calculate Spearman's rank correlation coefficient (RS).

C. General Program

```
C
C
C     ***********************************************************************
C     * THIS PROGRAM CALCULATES THE SPEARMAN RANK CORRELATION COEFFI-    *
C     * CIENT.                                                           *
C     ***********************************************************************
C
C
C
C     ***********************************************************************
C     * STEP 1.  DETERMINE THE NUMBER (N) OF PAIRS OF RANKS (X, Y) AND   *
C     * INSURE THAT THIS NUMBER IS 2 OR GREATER.                         *
C     ***********************************************************************
C
      READ (5, 500) N
  500 FORMAT (I6)
      IF (N .LT. 2) STOP
C
C     ***********************************************************************
C     * STEP 2.  ACCUMULATE THE SUM OF THE SQUARED DIFFERENCES IN RANK   *
C     * (SSQDIF).                                                        *
C     ***********************************************************************
C
C
C
C     ***********************************************************************
C     * INITIALIZE THE ACCUMULATOR OF THE SUM OF THE SQUARED DIFFERENCES *
C     * IN RANK (SSQDIF).                                                *
C     ***********************************************************************
C
      SSQDIF = 0.0
C
C     ***********************************************************************
C     * REPETITIVELY CYCLE THROUGH THE SET OF INSTRUCTIONS FROM THE ONE  *
C     * FOLLOWING THE DO STATEMENT THROUGH AND INCLUDING THE STATEMENT   *
C     * NUMBERED 25.  THIS SET OF INSTRUCTIONS WILL BE CYCLED THROUGH N  *
C     * TIMES.                                                           *
C     ***********************************************************************
C
      DO 25 I = 1, N
C
C     ***********************************************************************
C     * GET THE NEXT PAIR OF RANKS (X, Y).                               *
C     ***********************************************************************
C
      READ (5, 501) X, Y
  501 FORMAT (2F10.3)
C
C     ***********************************************************************
C     * ACCUMULATE THE SUM OF THE SQUARED DIFFERENCES IN RANK (SSQDIF).  *
```

```
C
C       ****************************************************************
        SSQDIF = SSQDIF + ((X - Y) ** 2)
   25   CONTINUE
C
C
C       ****************************************************************
C       * STEP 3.  CALCULATE THE SPEARMAN RANK CORRELATION COEFFICIENT  *
C       * (RS).                                                         *
C       ****************************************************************
C

        RS = 1.0 - ((6.0 * SSQDIF) / (N * ((N * N) - 1)))
        WRITE (6, 600) RS
  600   FORMAT (1X,45HTHE SPEARMAN RANK CORRELATION COEFFICIENT IS ,
       *     F10.3)
        STOP
        END
```

D. Example Problems

Example 1   Given the following ranked data:

X	Y
1	2
3	5
4	1
7	8
6	4
8	7

find the Spearman's rank correlation coefficient.

Input Data

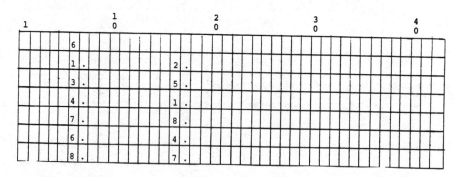

Solution   THE SPEARMAN RANK CORRELATION COEFFICIENT IS 0.429

Example 2   Two TV programs are rated by a consumer group as follows:

Program A	Program B
8	4
6	5
4	2
6	3
4	4
5	2

Find Spearman's rank correlation coefficient.

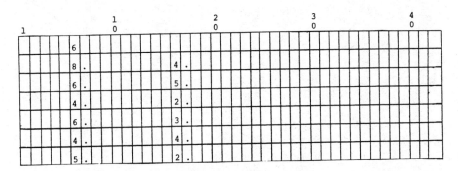

Input Data

THE SPEARMAN RANK CORRELATION COEFFICIENT IS −0.114   *Solution*

☆ Testing the Significance of Pearson's Product Moment Correlation Coefficient

A. Statement of Problem

Write a computer program which will:

(a) Compute Pearson's product correlation coefficient.

(b) Test whether the correlation coefficient is significantly different from zero or from a specified nonzero value. If $\rho = 0$ is to be tested against the alternative hypothesis $\rho \neq 0$, the test statistic is a t statistic, which is given as:

$$t = \frac{r}{\sqrt{\dfrac{1 - r^2}{n - 2}}}$$

where r is the correlation coefficient of a sample, n is the number of paired variates of the sample having $n - 2$

degrees of freedom. However, if the null hypothesis $\rho = \rho_0$, then our test statistic

$$z = \frac{Z - m_z}{\sigma_z}$$

satisfies an approximately normal distribution with:

$$Z = \tfrac{1}{2}\ln_e \frac{1 + r}{1 - r}$$

$$m_z = \tfrac{1}{2}\ln_e \frac{1 + \rho}{1 - \rho}$$

$$\sigma_z = \frac{1}{\sqrt{n - 3}}$$

where z is the standard normal random variable, Z is the Fisher's Z statistic, r is the correlation coefficient of the sample, ρ is the correlation coefficient of the population, n is the number of pairs of variates, and m_z and σ_z are the mean and standard deviation.

B. Algorithm

1. Calculate the correlation coefficient (R). This step is identical to the earlier program which computed the Pearson product moment correlation coefficient.

2. Determine the value (RHO) to which the calculated correlation coefficient (R) is to be compared.

3. If RHO is not equal to 0.0, go to Step 6. Otherwise, continue with the next step.

4. Determine if the correlation coefficient (R) is significantly different than 0.0. Calculate $T = R/\sqrt{(1 - R^2)/(N - 2)}$, where N is the number of pairs in the sample. Print the value of T. The task is complete.

5. Calculate $z = \dfrac{Z - m_z}{\sigma_z}$ and print z.

C. General Program

```
C
C    **********************************************************************
C    *  THIS PROGRAM TESTS THE SIGNIFICANCE OF A PEARSON PRODUCT MOMENT  *
C    *  CORRELATION COEFFICIENT.                                         *
```

```
C     ************************************************************************
C
C
C     ************************************************************************
C     * INFORM THE COMPUTER THAT MZ WILL BE A REAL VARIABLE RATHER THAN *
C     * AN INTEGER VARIABLE.                                            *
C     ************************************************************************
C
      REAL MZ
C
C     ************************************************************************
C     * STEP 1.  CALCULATE THE CORRELATION COEFFICIENT (R).   THIS FIRST *
C     * STEP IS IDENTICAL TO THE EARLIER PROGRAM THAT CALCULATED THE     *
C     * PEARSON PRODUCT MOMENT CORRELATION COEFFICIENT.                  *
C     ************************************************************************
C
      READ (5, 500) N
  500 FORMAT (I6)
      IF (N .LT. 2) STOP
      SUMX = 0.0
      SUMY = 0.0
      SUMX2 = 0.0
      SUMY2 = 0.0
      SUMXY = 0.0
      DO 10 I = 1, N
      READ (5, 501) X, Y
  501 FORMAT (2F10.3)
      SUMX = SUMX + X
      SUMY = SUMY + Y
      SUMX2 = SUMX2 + (X * X)
      SUMY2 = SUMY2 + (Y * Y)
      SUMXY = SUMXY + (X * Y)
   10 CONTINUE
      RNUM = (N * SUMXY) - (SUMX * SUMY)
      RDENOM = SQRT(((N * SUMX2 - (SUMX ** 2.0)) * (N * SUMY2 - (SUMY
     *   ** 2.0))))
      R = RNUM / RDENOM
      WRITE (6, 600) R
  600 FORMAT (1X,31HTHE CORRELATION COEFFICIENT IS , F10.3)
C
C     ************************************************************************
C     * STEP 2.  DETERMINE THE VALUE (RHO) TO WHICH THE CALCULATED COR- *
C     * RELATION COEFFICIENT (R) IS TO BE COMPARED.                     *
C     ************************************************************************
C
      READ (5, 502) RHO
  502 FORMAT (F10.3)
      WRITE (6, 601) RHO
  601 FORMAT (1X,20HTHE VALUE OF RHO IS , F10.3)
C
C     ************************************************************************
C     * STEP 3.  IF RHO IS NOT EQUAL TO 0.0 BRANCH TO THE Z TRANSFORMA- *
C     * TION ROUTINE; OTHERWISE CONTINUE TO THE NEXT INSTRUCTION.       *
C     ************************************************************************
C
      IF (RHO .NE. 0.0) GO TO 25
C
C     ************************************************************************
C     * STEP 4.  THE FOLLOWING ROUTINE DETERMINES WHETHER THE CORRELA-  *
C     * TION COEFFICIENT (R) IS SIGNIFICANTLY DIFFERENT THAN 0.0        *
C     ************************************************************************
C
C
C     ************************************************************************
C     * CALCULATE THE VALUE OF T AS INDICATED IN THE GIVEN FORMULA.     *
```

```
C
C     *****************************************************************
C
      T = R / SQRT(((1.0 - (R * R)) / (N - 2)))
      WRITE (6, 602) T
  602 FORMAT (1X,18HTHE VALUE OF T IS , F10.3)
      STOP
C
C
C     *****************************************************************
C     * STEP 5.  THE FOLLOWING ROUTINE DETERMINES WHETHER THE CORRELA- *
C     * TION COEFFICIENT (R) IS SIGNIFICANTLY DIFFERENT THAN SOME NON-  *
C     * ZERO VALUE.                                                     *
C     *****************************************************************
C
C
C
C     *****************************************************************
C     * CALCULATE THE VARIOUS COMPONENTS OF THE GIVEN FORMULA.          *
C     *****************************************************************
C
   25 CAPZ = ALOG((1.0 + R) / (1.0 - R)) / 2.0
      MZ = ALOG((1.0 + RHO) / (1.0 - RHO)) / 2.0
      SIGMAZ = 1.0 / SQRT(N - 3.0)
C
C
C     *****************************************************************
C     * CALCULATE THE VALUE OF Z AS INDICATED IN THE GIVEN FORMULA.     *
C     *****************************************************************
C
      Z = (CAPZ - MZ) / SIGMAZ
      WRITE (6, 603) Z
  603 FORMAT (1X,18HTHE VALUE OF Z IS , F10.3)
   40 STOP
      END
```

D. Example Problems

Example 1   The number of hours studied per week by a group of psychology students along with their accompanying final examination grades are recorded below:

Hours studied/week	Final grade
10	85
5	70
13	76
20	81
14	68
9	95

Find the Pearson's product moment correlation coefficient for the above data.

THE CORRELATION COEFFICIENT IS − 0.023   *Solution*

THE VALUE OF RHO IS 0.0

THE VALUE OF T IS − 0.045

The number of defective parts and the number of total parts produced   *Example 2*
are listed for seven production workers as:

Number of defective parts	Total parts produced
36	118
25	111
12	136
17	162
24	98
29	154
9	101

Find the Pearson's product moment correlation coefficient for the
above data.

Input Data

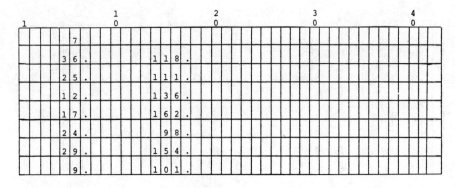

Solution   THE CORRELATION COEFFICIENT IS 0.019

THE VALUE OF RHO IS 0.0

THE VALUE OF T IS 0.042

1. Fit the following data to a linear regression model:

X	Y
12	22
10	18
15	24
18	32
17	25

2. The following data over a seven-year period indicated the number of lenses sold by Southern Optical of Greensboro, North Carolina. Fit the data to an exponential curve and determine Southern Optical's expected growth in sales of lenses for years 8–12.

Year	Number of lenses sold
1	15,000
2	18,000
3	30,200
4	55,000
5	61,000
6	75,800
7	85,000

3. Using the data in Problem 1, find the correlation coefficient of the given data.

4. Ten employees were rated on two sets of skills. Compute the correlation coefficient for the ten employees on the given two skill tests.

Employee	Rating for skill A	Rating for skill B
A	15	20
B	12	10
C	25	20
D	30	28
E	17	20
F	25	21
G	30	21
H	18	22
I	14	16
J	27	31

Chapter 8

Analysis of Variance

☆ **Analysis of Variance: One-way Classification**

A. Statement of Problem

Suppose that random samples each of size n_j are drawn from each of c populations with individual variates X_{ij}. Assume that the c populations are independent and normally distributed with means μ_1, μ_2, ..., μ_c and having a common variance σ^2. Let us write a program which:

(a) Computes the sum of squares as given by equations (1), (2), and (3) below:

$$SS_t = \sum_{i=1}^{c} \sum_{j=1}^{n_j} X_{ij}^2 - \frac{T^2..}{N} \tag{1}$$

(total sum of squares)

$$SS_c = \sum_{j=1}^{c} \left(\frac{T^2_{\cdot j\cdot}}{n_j} \right) - \frac{T^2_{\cdot \cdot \cdot}}{N} \tag{2}$$

(sum of squares of the column)

$$SS_e = SS_t - SS_c \tag{3}$$

(sum of squares of error)

where: $T_{\cdot \cdot}$ is the grand total.

$n_1, n_2, \ldots, n_c$ are the total number of variates within each sample.

$T_{j\cdot}$ are the column totals.

$$N = \sum_{j=1}^{c} n_j \text{ indicates the total variates within the population.}$$

(b)   Determines the ANOVA Table as indicated below.

Source of variation	Sum of squares	Degrees of freedom	Mean square	Compute F
Column Means	SS_c	$c - 1$	$s_1^2 = \dfrac{SS_c}{c - 1}$	$F = \dfrac{s_1^2}{s_2^2}$
Error (Residual)	SS_e	$N - c$	$s_2^2 = \dfrac{SS_e}{N - c}$	
Total	SS_t	$N - 1$		

B. Algorithm

1. Determine the number of samples (C) and insure it is greater than 1.

2. Perform the following for each sample.

    a. Determine the sample size (NJ) and insure it is greater than 1.

    b. Accumulate the sum (N) of the sample sizes (NJ).

    c. Read each observation (X) in the sample and accumulate their sum (SUMXJ). Accumulate the sum of their squares (SUMX2) along with the squared observations of all preceding samples.

    d. Accumulate the sum (SUMXJ2) of the sum of the observations for this sample (SUMXJ) squared divided by the sample size (NJ).

  e. Accumulate the sum (TDTDT) of the sum of the observations for the sample (SUMXJ).

3. Calculate SS_t (SST) by subtracting the sum of all observations (TDTDT) squared divided by the number of observations (N) from the sum of the squares of all observations (SUMX2).

4. Calculate SS_c (SSC) by subtracting the sum of all observations (TDTDT) squared divided by the number of observations (N) from the sum of the squared sample sizes divided by the sample sizes (SUMXJ2).

5. Derive SS_e (SSE) by subtracting SS_c from SS_t.

6. Derive the remaining elements of ANOVA Tables as indicated.

C. General Program

```
C
C
C      ******************************************************************
C      * THIS PROGRAM PERFORMS A ONE WAY ANALYSIS  OF VARIANCE FOR EQUAL *
C      * OR UNEQUAL SAMPLE SIZES.                                        *
C      ******************************************************************
C
       INTEGER DFCM, DFE, DFT, C
C
C      ******************************************************************
C      * STEP 1.  DETERMINE THE NUMBER (C) OF SAMPLES AND INSURE IT IS   *
C      * GREATER THAN 1.                                                 *
C      ******************************************************************
C
       READ (5, 500) C
 500   FORMAT (I6)
       IF (C .LT. 2) STOP
C
C      ******************************************************************
C      * STEP 2.  SET UP THE LOOP TO PROCESS ALL OBSERVATIONS.           *
C      ******************************************************************
C
C
C      ******************************************************************
C      * INITIALIZE THE VARIOUS ACCUMULATORS.                           *
C      ******************************************************************
C
       N = 0
       TDTDT = 0.0
       SUMX2 = 0.0
       SUMXJ2 = 0.0
       DO 10 J = 1, C
C
C      ******************************************************************
C      * STEP 2(A).  READ THE NEXT SAMPLE SIZE (NJ) AND INSURE IT IS     *
C      * GREATER THAN 1.                                                 *
C      ******************************************************************
C
       READ (5, 501) NJ
 501   FORMAT (I6)
       IF (NJ .LT. 2) STOP
C
C      ******************************************************************
C      * INITIALIZE THE SUM OF THE OBSERVATIONS (SUMXJ) FOR THIS GROUP.  *
```

```
C     ****************************************************************
C
      SUMXJ = 0.0
C
C     ****************************************************************
C     * STEP 2(B).  ACCUMULATE THE SUM (N) OF THE SAMPLE SIZES.      *
C     ****************************************************************
C
      N = N + NJ
C
C     ****************************************************************
C     * STEP 2(C). READ IN THE OBSERVATIONS (X) AND ACCUMULATE THEIR SUM*
C     * (SUMXJ) AND THE SUM OF THEIR SQUARES (SUMX2).               *
C     ****************************************************************
C
      DO 20 I = 1, NJ
      READ (5, 502) X
  502 FORMAT (F10.3)
      SUMXJ = SUMXJ + X
   20 SUMX2 = SUMX2 + (X * X)
C
C     ****************************************************************
C     * STEP 2(D).  ACCUMULATE THE SUM (SUMXJ2) OF THE SUM OF THE OBS- *
C     * ERVATIONS FOR THIS SAMPLE (SUMXJ) SQUARED DIVIDED BY THE SAMPLE *
C     * SIZE (NJ).                                                  *
C     ****************************************************************
C
      SUMXJ2 = SUMXJ2 + ((SUMXJ * SUMXJ) / NJ)
C
C     ****************************************************************
C     * STEP 2(E).  ACCUMULATE THE SUM (TOTOT) OF THE SUM OF THE OBS- *
C     * ERVATIONS (SUMXJ).                                          *
C     ****************************************************************
C
   10 TOTOT = TOTOT + SUMXJ
C
C     ****************************************************************
C     * STEP 3.  CALCULATE SST.                                     *
C     ****************************************************************
C
      SST = SUMX2 - ((TOTOT * TOTOT) / N)
C
C     ****************************************************************
C     * STEP 4.  CALCULATE SSC.                                     *
C     ****************************************************************
C
      SSC = SUMXJ2 - ((TOTOT * TOTOT) / N)
C
C     ****************************************************************
C     * STEP 5.  CALCULATE SSE.                                     *
C     ****************************************************************
C
      SSE = SST - SSC
C
C     ****************************************************************
C     * STEP 6.  CALCULATE THE REMAINING ELEMENTS OF THE ANOVA TABLE. *
C     ****************************************************************
C
      DFCM = C - 1
      DFE = N - C
      DFT = N - 1
      S1SQ = SSC / DFCM
      S2SQ = SSE / DFE
      F = S1SQ / S2SQ
      WRITE (6, 600)
```

```
600   FORMAT (1X,54HSOURCE OF VARIATION   SUM OF SQUARES      DF MEAN SQUAR
     *ES, 9X, 1HF )
      WRITE (6, 601) SSC, DFCM, S1SQ, F
601   FORMAT (1X,23HCOLUMN MEANS               , F10.3, 2X, I6, 2X, F10.3,
     *   1X, F10.3)
      WRITE (6, 602) SSE, DFE, S2SQ
602   FORMAT (1X,23HERROR                      , F10.3, 2X, I6, 2X, F10.3)
      WRITE (6, 603) SST, DFT
603   FORMAT (1X, 23HTOTAL                     , F10.3, 2X, I6)
      STOP
      END
```

D. Example Problems

Example 1   The following measurements are obtained for five groups of subjects on a psychomotor test:

I	II	III	IV	V
5	12	15	16	17
8	8	22	15	19
8	14	24	26	28
10	7	28	20	16
12	16	20	18	14

Find:

(a) The sums of squares.

(b) An ANOVA Table for a one-way analysis of variance.

Input Data

```
     1           1           2           3           4
                 0           0           0           0
         5
         5
         5 .
         8 .
         8 .
     1   0 .
     1   2 .
         5
     1   2 .
         8 .
     1   4 .
         7 .
     1   6 .
         5
     1   5 .
     2   2 .
     2   4 .
     2   8 .
     2   0 .
         5
     1   6 .
     1   5 .
     2   6 .
     2   0 .
     1   8 .
         5
     1   7 .
     1   9 .
     2   8 .
     1   6 .
     1   4 .
```

Solution

SOURCE OF VARIATION	SUM OF SQUARES	DF	MEAN SQUARE	F
COLUMN MEANS	631.840	4	157.960	8.447
ERROR	374.004	20	18.700	
TOTAL	1005.844	24		

Example 2   An experiment was performed on a number of psychology students and involved their ability to recall nonsense words. Four methods of presentation were used. The results of the number of words recalled by each method of presentation are indicated below:

I	II	III	IV
4	8	7	13
3	7	5	11
6	5	6	10
5	10	12	14
	8	4	9
	6		12

Find:

(a) The sums of squares.

(b) An ANOVA Table for a one-way analysis of variance.

Input Data

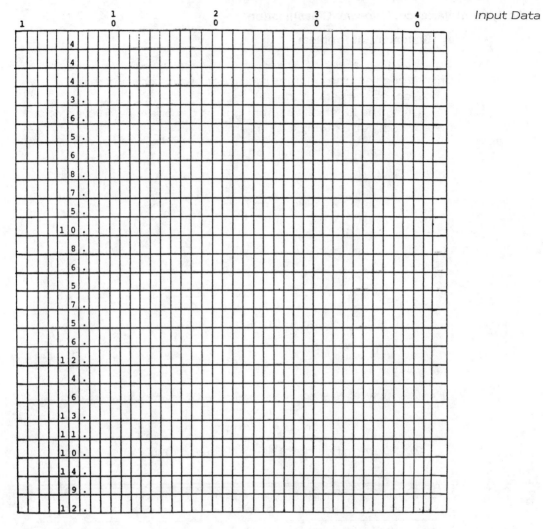

SOURCE OF VARIATION	SUM OF SQUARES	DF	MEAN SQUARE	F
COLUMN MEANS	131.938	3	43.979	9.756
ERROR	76.634	17	4.508	
TOTAL	208.572	20		

Solution

☆ Analysis of Variance: Two-way Classification

A. Statement of Problem

Suppose that a set of observations is classified according to two criteria and that the columns represent one criteria and that the rows represent the other criteria.

Using the identical assumptions of the one-way analysis of variance model, let us write a program which:

(a) Computes the sum of squares:

$$SS_t = \sum_{i=1}^{r} \sum_{j=1}^{c} X_{ij}^2 - \frac{T^2..}{N} \tag{1}$$

$$SS_r = \frac{\sum_{i=1}^{r} T_i^2.}{c} - \frac{T^2..}{N} \tag{2}$$

$$SS_c = \frac{\sum_{j=1}^{c} T._j^2}{r} - \frac{T^2..}{N} \tag{3}$$

$$SS_e = SS_t - SS_r - SS_c \tag{4}$$

where: $N = rc$.
T_i. are the row totals.
$T._j$ are the column totals.

(b) Determines the ANOVA Table as indicated below:

Source of variation	Sum of squares	Degrees of freedom	Mean square	Compute F
Row means	SS_r	$r - 1$	$s_1^2 = \dfrac{SS_r}{r - 1}$	$F_1 = \dfrac{s_1^2}{s_3^2}$
Column means	SS_c	$c - 1$	$s_2^2 = \dfrac{SS_c}{c - 1}$	$F_2 = \dfrac{s_2^2}{s_3^2}$
Error (residual)	SS_e	$(r - 1)(c - 1)$	$s_3^2 = \dfrac{SS_e}{(r - 1)(c - 1)}$	
Total	SS_t	$rc - 1$		

B. Algorithm

1. Determine the number of rows (R) and the number of columns (C); insure that the number of columns is greater than 1 and equal to or less than 100. Insure that the number of rows is at least 2.

2. Perform the following for *each* row of data:

   a. Read all C observations (X) in the row.

   b. Accumulate the sum of the squared values in the row (SUMX2).

   c. Accumulate the sum of the values in the row (SUMXI).

   d. Accumulate the sums (SUMXJ(J)) of the values *by column*.

   e. Accumulate the sum (TDTDT) of the sum of the observations for the row.

   f. Accumulate the sum (SUMXI2) of the row totals (SUMXI) squared divided by the number of columns (C).

3. Sum the column totals (SUMXJ(J)) squared divided by the number of rows (R). Store the sum in (SUMXJ2).

4. Calculate SS_t (SST) by subtracting the sum of the observations (TDTDT) squared divided by the number of observations from the sum of the squares of the observations (SUMX2).

5. Calculate SS_r (SSR) by subtracting the sum of the observations (TDTDT) squared divided by the sum of the row totals squared divided by the number of columns (C).

6. Calculate SS_c (SSC) by subtracting the sum of the observations (TDTDT) squared divided by the number of observations from the sum of the column totals squared divided by the number of rows (SUMXJ2).

7. Calculate SS_e (SSE) by subtracting SSR and SSC from SST.

8. Calculate the remaining elements of the ANOVA Table as indicated.

C. General Program

```
C
C    ***********************************************************************
C    *  THIS PROGRAM PERFORMS A 2 WAY ANALYSIS OF VARIANCE.                *
C    ***********************************************************************
C
      INTEGER DFRM, DFCM, DFE, DFT, R, C
      DIMENSION SUMXJ(100), X(100)
C
C    ***********************************************************************
C    *  STEP 1.  DETERMINE THE NUMBER (R) OF ROWS AND THE NUMBER (C) OF    *
C    *  COLUMNS.                                                           *
```

```
C
C      ********************************************************************
C
       READ (5, 500) R, C
  500  FORMAT (2I6)
       IF (C .LT. 2 .OR. C .GT. 100) STOP
       IF (R .LT. 2) STOP
C
C      ********************************************************************
C      * STEP 2.  PROCESS EACH ROW OF DATA.                              *
C      ********************************************************************
C
C
C      ********************************************************************
C      * INITIALIZE THE VARIOUS ACCUMULATORS.                            *
C      ********************************************************************
C
       TOTDT = 0.0
       SUMX2 = 0.0
       SUMXI2 = 0.0
C
C      ********************************************************************
C      * INITIALIZE THE ARRAY (SUMXJ) USED TO ACCUMULATE THE COLUMN TO-  *
C      * TALS.                                                           *
C      ********************************************************************
C
       DO 5 J = 1, C
    5  SUMXJ(J) = 0.0
C
C      ********************************************************************
C      * SET UP THE LOOP TO READ ALL ROWS OF DATA.                       *
C      ********************************************************************
C
       DO 10 I = 1, R
C
C      ********************************************************************
C      * INITIALIZE THE ACCUMULATOR OF THE ROW TOTAL.                    *
C      ********************************************************************
C
       SUMXI = 0.0
C
C      ********************************************************************
C      * STEP 2(A).  READ IN ONE ROW OF DATA.                           *
C      ********************************************************************
C
       READ (5, 501)(X(J), J = 1, C)
  501  FORMAT (8F10.3)
       DO 20 J = 1, C
C
C      ********************************************************************
C      * STEP 2(B).  ACCUMULATE THE SUM OF THE SQUARES (SUMX2).          *
C      ********************************************************************
C
       SUMX2 = SUMX2 + (X(J) * X(J))
C
C      ********************************************************************
C      * STEP 2(C).  ACCUMULATE THE SUM OF THE ROW (SUMXI).              *
C      ********************************************************************
C
       SUMXI = SUMXI + X(J)
C
C      ********************************************************************
C      * STEP 2(D).  ACCUMULATE THE SUMS BY COLUMN (SUMXJ(J)).           *
C      ********************************************************************
C
   20  SUMXJ(J) = SUMXJ(J) + X(J)
```

```
C
C      ************************************************************
C      * STEP 2(E).  ACCUMULATE THE SUM (TOTDT) OF THE SUM OF THE ROWS.  *
C      ************************************************************
C
       TOTDT = TOTDT + SUMXI
C
C      ************************************************************
C      * STEP 2(F).  ACCUMULATE THE SUM (SUMXI2) OF THE ROW TOTALS      *
C      * SQUARED DIVIDED BY THE NUMBER OF COLUMNS (C).                  *
C      ************************************************************
C
   10  SUMXI2 = SUMXI2 + ((SUMXI * SUMXI) / C)
C
C      ************************************************************
C      * STEP 3.  ACCUMULATE THE SUM (SUMXJ2) OF THE COLUMN TOTALS      *
C      * SQUARED DIVIDED BY THE NUMBER OF ROWS (R).                    *
C      ************************************************************
C
C
C      ************************************************************
C      * INITIALIZE THE ACCUMULATOR (SUMXJ2).                          *
C      ************************************************************
C
       SUMXJ2 = 0.0
C
C      ************************************************************
C      * SET UP THE LOOP TO PROCESS ALL COLUMN TOTALS.                 *
C      ************************************************************
C
       DO 30 J = 1, C
   30  SUMXJ2 = SUMXJ2 + (SUMXJ(J) * SUMXJ(J) / R)
C
C      ************************************************************
C      * STEP 4.  CALCULATE SST.                                       *
C      ************************************************************
C
       SST = SUMX2 - ((TOTDT * TOTDT) / (C * R))
C
C      ************************************************************
C      * STEP 5.  CALCULATE SSR.                                       *
C      ************************************************************
C
       SSR = SUMXI2 - ((TOTDT * TOTDT) / (C * R))
C
C      ************************************************************
C      * STEP 6.  CALCULATE SSC.                                       *
C      ************************************************************
C
       SSC = SUMXJ2 - ((TOTDT * TOTDT) / (C * R))
C
C      ************************************************************
C      * STEP 7.  CALCULATE SSE.                                       *
C      ************************************************************
C
       SSE = SST - SSR - SSC
C
C      ************************************************************
C      * STEP 8.  CALCULATE THE REMAINING ELEMENTS OF THE ANOVA TABLE.  *
C      ************************************************************
C
       DFRM = R - 1
       DFCM = C - 1
       DFE = (R - 1) * (C - 1)
       DFT = R * C - 1
       S1SQ = SSR / DFRM
```

```
        S2SQ = SSC / DFCM
        S3SQ = SSE / DFE
        F1 = S1SQ / S3SQ
        F2 = S2SQ / S3SQ
        WRITE (6, 600)
600     FORMAT (1X,54HSOURCE OF VARIATION   SUM OF SQUARES      DF MEAN SQUAR
        WRITE (6, 601) SSR, DFRM, S1SQ, F1
601     FORMAT (1X,23HROW MEANS                  , F10.3, 2X, I6, 2X, F10.3,
    *      1X, F10.3)
        WRITE (6, 602) SSC, DFCM, S2SQ, F2
602     FORMAT (1X,23HCOLUMN MEANS               , F10.3, 2X, I6, 2X, F10.3,
    *      1X, F10.3)
        WRITE (6, 603) SSE, DFE, S3SQ
603     FORMAT (1X,23HERROR                      , F10.3, 2X, I6, 2X, F10.3)
        WRITE (6, 604) SST, DFT
604     FORMAT (1X,23HTOTAL                      , F10.3, 2X, I6)
        STOP
C       END
```

D. Example Problems

Example 1   Three varieties of corn are used in an experiment utilizing four different types of fertilizer treatments. The productivity in bushels per acre is indicated below:

Fertilizer variety	I	II	III
1	40	35	41
2	64	59	72
3	42	40	36
4	38	37	31

Find:

(a) The sums of squares.

(b) An ANOVA Table for a two-way analysis of variance.

Input Data

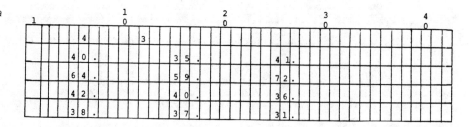

Solution

SOURCE OF VARIATION	SUM OF SQUARES	DF	MEAN SQUARE	F
ROW MEANS	2274.914	3	758.305	7.753
COLUMN MEANS	87.168	2	43.584	0.446
ERROR	586.836	6	97.806	
TOTAL	2948.918	11		

Example 2

The following scores represent the final grades received by six students in statistics, botany, history, and psychology:

Student	Statistics	Botany	History	Psychology
1	70	55	93	75
2	74	68	89	79
3	82	73	87	65
4	85	84	86	81
5	96	88	92	95
6	77	71	84	73

Find:

(a) The sums of squares.

(b) An ANOVA Table for a two-way analysis of variance.

Input Data

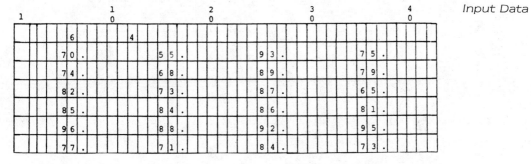

Solution

SOURCE OF VARIATION	SUM OF SQUARES	DF	MEAN SQUARE	F
ROW MEANS	1019.875	5	203.975	4.568
COLUMN MEANS	740.187	3	246.729	5.525
ERROR	669.812	15	44.654	
TOTAL	2429.875	23		

☆ Analysis of Variance: Two-way Classification with Replication

A. Statement of Problem

Given a rectangular array consisting of r rows and c columns; each of the rc cells contains n observations. In general, X_{ijk} refers to an element in the ith row, jth column, and kth replicate. Write a program which:

(a) Computes the sums of squares below:

$$SS_t = \sum_{i=1}^{r} \sum_{j=1}^{c} \sum_{k=1}^{n} X_{ijk}^2 - \frac{T^2 \dots}{rcn} \tag{1}$$

$$SS_r = \sum_{i=1}^{r} \frac{T_{i..}^2}{cn} - \frac{T^2 \dots}{rcn} \tag{2}$$

$$SS_c = \sum_{i=1}^{c} \frac{T_{.j.}^2}{rn} - \frac{T^2 \dots}{rcn} \tag{3}$$

$$SS_{rc} = \frac{\sum_{i=1}^{r} \sum_{j=1}^{c} T_{ij.}^2}{n} - \frac{\sum_{i=1}^{r} T_{i..}^2}{cn} - \frac{\sum_{j=1}^{c} T_{.j.}^2}{rn} + \frac{T^2 \dots}{rcn} \tag{4}$$

$$SS_e = SS_t - SS_r - SS_c - SS_{rc} \tag{5}$$

(b) Determines an ANOVA Table as indicated below:

Source of variation	Sum of squares	Degrees of freedom	Mean square	Calculated F
Row means	SS_r	$r - 1$	$s_1^2 = \dfrac{SS_r}{r-1}$	$F_1 = \dfrac{s_1^2}{s_4^2}$
Column means	SS_c	$c - 1$	$s_2^2 = \dfrac{SS_c}{c-1}$	$F_2 = \dfrac{s_2^2}{s_4^2}$
Interaction	SS_{rc}	$(r-1)(c-1)$	$s_3^2 = \dfrac{SS_{rc}}{(r-1)(c-1)}$	$F_3 = \dfrac{s_3^2}{s_4^2}$
Error (residual)	SS_e	$rc(n-1)$	$s_4^2 = \dfrac{SS_e}{rc(n-1)}$	
Total	SS_t	$rcn - 1$		

B. Algorithm

1. Determine the number of rows (R), the number of columns (C), and the number of replicates (N). The number of columns must be greater than 1 and equal to or less than 100. The number of rows and replicates must both be greater than one.

2. Do the following for *each* row of cells.

    a. Do the following for *each* of the *n* rows of replicates (OBS(J)):

        (1) Accumulate the sum of observations (T3DT) for this and all preceding rows of replicates.

        (2) Accumulate the sum of the squares of the observations (SUMO2) for this and all preceding rows of replicates.

        (3) Accumulate the sum of the replicates in a cell (SUMCEL(J)) for each column.

    b. Accumulate the sum of the observations in the row (SUMOI).

    c. Accumulate the sum of squares (SUMCL2) of the sum of the replicates in each cell (SUMCEL(J)) divided by the number of replicates (N). This sum is of current row and all previous rows.

    d. Accumulate the sums (SUMOJ(J)) of the sums of the replicates in a column (SUMCEL(J)). These sums are for the current and all preceding rows.

    e. Accumulate the sum (SUMOI2) of the squares of the sum of observations in a row (SUMOI) divided by the number of columns (C) times the number of replicates (N).

3. Accumulate the sum (SUMOJ2) of the squares of the sums (SUMOJ(J)) of the sums of the replicates in a column.

4. Calculate SS_t (SST) by subtracting the square of the sum of the observations (T3DT) divided by the number of rows (R) times the number of columns (C) times the number of replicates (N) from the sum of the squares of the observations (SUMO2).

5. Calculate SS_r (SSR) by subtracting the square of the sum of the observations (T3DT) divided by the number of rows (R) times the number of columns (C) times the number of replicates (N) from the variable (SUMOI2).

6. Calculate SS_c (SSC) by subtracting the square of (T3DT) divided by $R \times C \times N$ from the variable (SUMOJ2).

7. Calculate SS_{rc} (SSRC) by adding the variable (SUMCL2) to the term (T3DT) squared divided by $R \times C \times N$ and then subtracting (SUMOJ2) and (SUMOI2).

8. Calculate SS_e (SSE) by subtracting SS_{rc}, SS_c, and SS_r from SS_t

9. Calculate the remaining elements of the ANOVA Table as indicated.

C. General Program

```
C
C
C       ********************************************************************
C       * THIS PROGRAM PERFORMS AN ANALYSIS OF VARIANCE FOR A TWO WAY      *
C       * CLASSIFICATION WITH REPLICATION.                                 *
C       ********************************************************************
C
        INTEGER R, C, DFRM, DFRC, DFE, DFT
        DIMENSION OBS(100), SUMOJ(100), SUMCEL(100)
C
C       ********************************************************************
C       * STEP 1.  DETERMINE THE NUMBER OF ROWS (R), THE NUMBER OF COLUMNS*
C       * (C) AND THE NUMBER OF REPLICATES.  INSURE THAT THESE VALUES ARE *
C       * WITHIN THE PROPER RANGES.                                        *
C       ********************************************************************
C
        READ (5, 500) R, C, N
  500   FORMAT (3I6)
        IF (C .LT. 2 .OR. C .GT. 100) STOP
        IF (R .LT. 2) STOP
        IF (N .LT. 2) STOP
C
C       ********************************************************************
C       * STEP 2.  SET UP THE PROCESSING FOR EACH ROW OF CELLS.           *
C       ********************************************************************
C
C
C       ********************************************************************
C       * INITIALIZE THE VARIOUS ACCUMULATORS.                            *
C       ********************************************************************
C
        T3DT = 0.0
        SUMO2 = 0.0
        SUMCL2 = 0.0
        SUMOI2 = 0.0
C
C       ********************************************************************
C       * INITIALIZE THE FIRST C ELEMENTS OF THE ARRAY SUMOJ.            *
C       ********************************************************************
C
        DO 5 J = 1, C
    5   SUMOJ(J) = 0.0
C
C       ********************************************************************
C       * SET UP THE LOOP TO PROCESS ALL R ROWS OF CELLS.                *
C       ********************************************************************
C
        DO 30 I = 1, R
C
C       ********************************************************************
C       * INITIALIZE THE FIRST C ELEMENTS OF THE ARRAY SUMCEL.           *
```

```
C
C      ****************************************************************
       DO 7 J = 1, C
  7    SUMCEL(J) = 0.0
C
C      ****************************************************************
C      * STEP 2(A).SET UP THE LOOP TO PROCESS ALL N ROWS OF REPLICATES. *
C      ****************************************************************
C
       DO 10 K = 1, N
C
C      ****************************************************************
C      * READ IN THE J OBSERVATIONS IN THIS ROW.                     *
C      ****************************************************************
C
       READ (5, 501) (OBS(J), J = 1, C)
 501   FORMAT (8F10.3)
C
C      ****************************************************************
C      * SET UP THE LOOP TO PROCESS THE J OBSERVATIONS IN THIS ROW.  *
C      ****************************************************************
C
       DO 10 J = 1, C
C
C      ****************************************************************
C      * STEP 2(A.1).  ACCUMULATE THE SUM OF THE OBSERVATIONS (T3DT).  *
C      ****************************************************************
C
       T3DT = T3DT + OBS(J)
C
C      ****************************************************************
C      * STEP 2(A.2).  ACCUMULATE THE SUM OF THE SQUARES OF THE OBSERVA- *
C      * TIONS (SUMO2).                                              *
C      ****************************************************************
C
       SUMO2 = SUMO2 + (OBS(J) * OBS(J))
C
C      ****************************************************************
C      * STEP 2(A.3).  ACCUMULATE THE SUM OF THE REPLICATES IN A CELL  *
C      * (SUMCEL(J)) FOR EACH COLUMN.                                *
C      ****************************************************************
C
 10    SUMCEL (J) = SUMCEL(J) + OBS(J)
C
C      ****************************************************************
C      * STEP 2(B).  ACCUMULATE THE SUM OF THE OBSERVATIONS IN THIS ROW *
C      * OF CELLS (SUMOI).                                           *
C      ****************************************************************
C
C
C      ****************************************************************
C      * INITIALIZE THE ACCUMULATOR OF THE SUM OF THE OBSERVATIONS IN A *
C      * ROW (SUMOI).                                                *
C      ****************************************************************
C
       SUMOI = 0.0
C
C      ****************************************************************
C      * SET UP THE LOOP TO SUM THE C CELL SUMS.                     *
C      ****************************************************************
C
       DO 15 J = 1, C
C
C      ****************************************************************
C      * ACCUMULATE THE SUM FOR THIS ROW OF CELLS.                   *
```

```
C     ***********************************************************************
C
C         SUMOI = SUMOI + SUMCEL(J)
C
C     ***********************************************************************
C     * STEP 2(C).  ACCUMULATE THE SUM OF THE SQUARES (SUMCL2) OF THE       *
C     * SUM OF THE REPLICATES IN EACH CELL DIVIDED BY THE NUMBER OF         *
C     * REPLICATES.                                                         *
C     ***********************************************************************
C
C         SUMCL2 = SUMCL2 + ((SUMCEL(J) * SUMCEL(J)))/ N
C
C     ***********************************************************************
C     * STEP 2(D).  ACCUMULATE THE SUMS (SUMOJ(J)) OF THE SUMS OF THE       *
C     * REPLICATES IN A COLUMN.                                             *
C     ***********************************************************************
C
   15     SUMOJ(J) = SUMOJ(J) + SUMCEL(J)
C
C     ***********************************************************************
C     * STEP 2(E).  ACCUMULATE THE SUM (SUMOI2) OF THE SQUARES OF THE       *
C     * SUM OF THE OBSERVATIONS IN A ROW (SUMOI) DIVIDED BY THE NUMBER      *
C     * OF COLUMNS TIMES THE NUMBER OF REPLICATES.                          *
C     ***********************************************************************
C
   30     SUMOI2 = SUMOI2 + ((SUMOI * SUMOI) / (C * N))
C
C     ***********************************************************************
C     * STEP 3.  ACCUMULATE THE SUM (SUMOJ2) OF THE SQUARES OF THE SUMS     *
C     * (SUMOJ(J)) OF THE SUMS OF THE REPLICATES IN A COLUMN.               *
C     ***********************************************************************
C
C
C
C     ***********************************************************************
C     * INITIALIZE THE ACCUMULATOR (SUMOJ2).                                *
C     ***********************************************************************
C
          SUMOJ2 = 0.0
C
C     ***********************************************************************
C     * SET UP THE LOOP TO PROCESS ALL C COLUMNS.                           *
C     ***********************************************************************
C
          DO 35 J = 1, C
C
C     ***********************************************************************
C     * ACCUMULATE THE SUM.                                                 *
C     ***********************************************************************
C
   35     SUMOJ2 = SUMOJ2 + ((SUMOJ(J) * SUMOJ(J)) / (R * N))
C
C     ***********************************************************************
C     * STEP 4.  CALCULATE SST.                                             *
C     ***********************************************************************
C
          SST = SUMO2 - ((T3DT * T3DT) / (R * C * N))
C
C     ***********************************************************************
C     * STEP 5.  CALCULATE SSR.                                             *
C     ***********************************************************************
C
          SSR = SUMOI2 - ((T3DT * T3DT) / (R * C * N))
C
C     ***********************************************************************
C     * STEP 6.  CALCULATE SSC.                                             *
```

```
C     ****************************************************************
C
C        SSC = SUMOJ2 - ((T3DT * T3DT) / (R * C * N))
C
C     ****************************************************************
C     * STEP 7.   CALCULATE SSRC.                                   *
C     ****************************************************************
C
C        SSRC = SUMCL2 - SUMOJ2 - SUMOI2 + ((T3DT * T3DT) / (R * C * N))
C
C     ****************************************************************
C     * STEP 8.   CALCULATE SSE.                                    *
C     ****************************************************************
C
C        SSE = SST - SSR - SSC - SSRC
C
C     ****************************************************************
C     * STEP 9.   CALCULATE THE REMAINING ELEMENTS OF THE ANOVA TABLE.  *
C     ****************************************************************
C
      DFRM = R - 1
      DFCM = C - 1
      DFRC = (R - 1) * (C - 1)
      DFE = (R * C) * (N - 1)
      DFT = R * C * N - 1
      S1SQ = SSR / DFRM
      S2SQ = SSC / DFCM
      S3SQ = SSRC / DFRC
      S4SQ = SSE / DFE
      F1 = S1SQ / S4SQ
      F2 = S2SQ / S4SQ
      F3 = S3SQ / S4SQ
      WRITE (6, 600)
  600 FORMAT (1X,54HSOURCE OF VARIATION  SUM OF SQUARES    DF MEAN SQUAR
     *ES, 9X, 1HF)
      WRITE (6, 601) SSR, DFRM, S1SQ, F1
  601 FORMAT (1X,23HROW MEANS              , F10.3, 2X, I6, 2X, F10.3,
     *  1X, F10.3)
      WRITE (6, 602) SSC, DFCM, S2SQ, F2
  602 FORMAT (1X,23HCOLUMN MEANS           , F10.3, 2X, I6, 2X, F10.3,
     *  1X, F10.3)
      WRITE (6, 603) SSRC, DFRC, S3SQ, F3
  603 FORMAT (1X,23HINTERACTION            , F10.3, 2X, I6, 2X, F10.3,
     *  1X, F10.3)
      WRITE (6, 604) SSE, DFE, S4SQ
  604 FORMAT (1X,23HERROR                  , F10.3, 2X, I6, 2X, F10.3)
      WRITE (6, 605) SST, DFT
  605 FORMAT (1X,23HTOTAL                  , F10.3, 2X, I6)
      STOP
      END
```

D. Example Problems

Suppose that we have three varieties of cucumbers planted on four   *Example 1*
identical plots of land with three different fertilizer treatments. The
yield in bushels per plot is as indicated:

Fertilizer	Variety of cucumbers		
	1	2	3
1	20 15 17 13	11 13 18 16	15 12 11 9
2	16 18 11 15	14 13 10 9	18 17 14 15
3	23 21 19 22	22 20 18 21	23 19 21 24

Find:

(a) The sums of squares.

(b) The ANOVA Table for a two-way analysis of variance model with replication.

Input Data

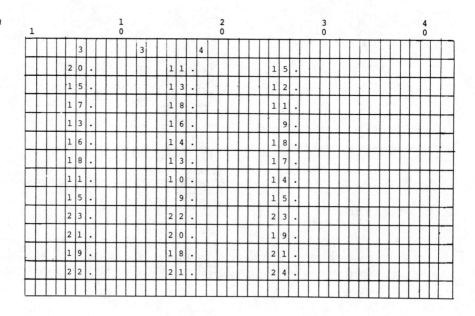

SOURCE OF VARIATION	SUM OF SQUARES	DF	MEAN SQUARE	F
ROW MEANS	382.719	2	191.359	32.342
COLUMN MEANS	26.055	2	13.027	2.202
INTERACTION	64,449	4	16.112	2.723
ERROR	159.750	27	5.917	
TOTAL	632.973	35		

The data in the following table represent the results of three quizzes   *Example 2*
obtained by six students in accounting, English, psychology, and
statistics:

Student	Accounting	English	Psychology	Statistics
1	80 68 74	71 70 73	81 80 79	67 69 71
2	71 73 75	84 81 83	73 78 82	70 65 71
3	82 63 74	65 82 74	81 80 75	60 63 59
4	88 92 87	92 86 91	94 90 93	87 91 98
5	75 81 87	78 73 74	73 76 82	55 62 71
6	91 82 75	81 91 85	80 81 79	48 52 59

Find:

(a) The sums of squares.

(b) The ANOVA Table for a two-way analysis of variance model with
replication.

Input Data

1	2	3	4
80.	71.	81.	67.
68.	70.	80.	69.
74.	73.	79.	71.
71.	84.	73.	70.
73.	81.	78.	65.
75.	83.	82.	71.
82.	65.	81.	60.
63.	82.	80.	63.
74.	74.	75.	59.
88.	92.	94.	87.
92.	86.	90.	91.
87.	91.	93.	98.
75.	78.	73.	55.
81.	73.	76.	62.
87.	74.	82.	71.
91.	81.	80.	48.
82.	91.	81.	52.
75.	85.	79.	59.

Solution

SOURCE OF VARIATION	SUM OF SQUARES	DF	MEAN SQUARE	F
ROW MEANS	2942.187	5	588.437	25.168
COLUMN MEANS	2028.875	3	676.292	28.926
INTERACTION	1657.687	15	110.512	4.727
ERROR	1122.250	48	23.380	
TOTAL	7751.000	71		

1. An advertising firm wished to determine the effectiveness of consumer response to a particular product and noted the following frequencies of consumer exposure for a seven-week duration:

Newspaper	Television	Radio
20	13	8
27	19	5
21	33	15
17	21	10
19	28	17
25	26	14
15	18	16

Using the above data, compute:

(a) The sums of squares.

(b) The appropriate ANOVA Table.

2. A market researcher noted the following program viewing preferences among these adult groups, as indicated below:

	20–30	30–40	40 and over
Comedy	80	75	43
Movies	105	110	95
Soap opera	45	68	73
Variety	70	85	113

Using the above data, find:

(a) The sums of squares.

(b) The ANOVA Table for this two-way design.

3. Three social planning areas reported the number of crimes in each of their respective areas for three years as follows:

SPA	Burglaries	Larcenies	Auto Thefts
1	20 15 17	24 27 17	31 26 22
2	31 32 25	28 19 26	45 42 38
3	21 16 13	12 15 9	20 18 16

Find:

(a) The total sums of squares.

(b) The ANOVA Table for this two-way design with replication.

Chapter 9
Nonparametric Tests

A. Statement of Problems

The runs test is a test for randomness. The null hypothesis which we wish to test is whether the observations that have been drawn from a parent population are random. Suppose the N_1 letters are of a given type and N_2 letters are of a second type. Furthermore, let us assume that r is the total number of runs (number of successive and similar letters) and that N_1 and N_2 are both greater than or equal to 20. Then the runs test is said to satisfy a normal distribution with:

Mean $\qquad \mu_r = \dfrac{2N_1N_2}{N_1 + N_2} + 1$

Standard Deviation $\sigma_r = \sqrt{\dfrac{2N_1N_2\,(2N_1N_2 - N_1 - N_2)}{(N_1 + N_2)^2\,(N_1 + N_2 - 1)}}$

Test Statistic $\qquad Z = \dfrac{r - \mu_r}{\sigma_r}$

where: μ_r is the mean number of runs.

σ_r is the standard deviation of the number of runs.

Write a computer program which will determine the Z test statistic for the runs test.

B. Algorithm

1. Perform the following steps for each observation:

   a. Read an observation (X).

   b. Each observation must be coded 1 or 2. A value other than 1 or 2 is used to indicate the end of the input data and that the program should proceed to Step 2.

   c. Count the number of observations coded 1 and the number of observations coded 2.

   d. If the current observation is not equal to the preceding observation, it indicates the beginning of another run. Count the number of runs (R).

2. Insure that the number of observations coded 1 (N1) and the number of observations coded 2 (N2) are both greater than or equal to 20.

3. Calculate the mean (UR) and standard deviation (SIGMAR) as indicated in the given formula.

4. Calculate the test statistic (Z) as indicated by the formula.

C. General Program

```
C
C
C     ****************************************************************
C     * THIS PROGRAM PERFORMS A TEST OF RANDOMNESS (THE RUNS TEST).  IT *
C     * ASSUMES THAT THE INPUT OBSERVATIONS ARE CODED 1 OR 2.         *
C     ****************************************************************
C
      DIMENSION OBS (100,2)
      INTEGER R, X
      N1 = 0
      N2 = 0
      R = 0
C
C
C     ****************************************************************
C     * THE VARIABLE LAST WILL BE USED TO STORE THE INITIAL NUMBER (A 1 *
C     * OR A 2) OF ANY RUN.  SUCCESSIVE OBSERVATIONS WILL BE READ AND   *
C     * COMPARED TO THIS VARIABLE.  AN OBSERVATION NOT EQUAL TO THE VAR-*
C     * IABLE LAST INDICATES THE BEGINNING OF A NEW RUN.  IN ORDER FOR  *
C     * THE FIRST OBSERVATION TO BE UNEQUAL TO LAST, WE WILL INITIALIZE *
C     * LAST TO AN ARBITRARY NUMBER OTHER THAN 1 OR 2.                 *
C     ****************************************************************
C
      LAST = -9
C
C
C     ****************************************************************
C     * STEP 1.  DETERMINE THE NUMBER OF OBSERVATIONS CODED 1 (N1), THE *
```

```
C     * NUMBER OF OBSERVATIONS CODED 2 (N2) AND THE NUMBER OF RUNS (R). *
C     ********************************************************************
C
C
C     ********************************************************************
C     * STEP 1(A).  READ AN OBSERVATION (X).                            *
C     ********************************************************************
C
  10    READ (5, 500) X
 500    FORMAT (I2)
C
C     ********************************************************************
C     * STEP 1(B).  AN OBSERVATION CODED OTHER THAN 1 OR 2 INDICATES THE*
C     * END OF THE INPUT DATA.  IF THERE IS NO MORE INPUT DATA BRANCH TO*
C     * STEP 2.                                                         *
C     ********************************************************************
C
        IF (X .NE. 1 .AND. X .NE. 2) GO TO 20
C
C     ********************************************************************
C     * STEP 1(C).  COUNT THE NUMBER OF OBSERVATIONS CODED 1 (N1) AND    *
C     * THE NUMBER OF OBSERVATIONS CODED 2 (N2).                        *
C     ********************************************************************
C
        IF (X .EQ. 1) N1 = N1 + 1
        IF (X .EQ. 2) N2 = N2 + 1
C
C     ********************************************************************
C     * STEP 1(D).  IF THIS OBSERVATION (X) EQUALS THE VARIABLE LAST GO *
C     * BACK AND READ ANOTHER OBSERVATION.  OTHERWISE CONTINUE TO THE   *
C     * NEXT SEQUENTIAL INSTRUCTION.                                    *
C     ********************************************************************
C
        IF (X .EQ. LAST) GO TO 10
C
C     ********************************************************************
C     * AT THIS POINT WE KNOW A NEW RUN IS BEGINNING.  COUNT THE NUMBER *
C     * OF RUNS (R) AND STORE THE INITIAL OBSERVATION (X) OF THIS RUN IN*
C     * THE VARIABLE LAST.                                              *
C     ********************************************************************
C
        R = R + 1
        LAST = X
C
C     ********************************************************************
C     * GO BACK TO READ ANOTHER OBSERVATION.                           *
C     ********************************************************************
C
        GO TO 10
C
C     ********************************************************************
C     * STEP 2.  INSURE THAT N1 AND N2 ARE BOTH GREATER THAN 20.        *
C     ********************************************************************
C
  20    IF (N1 .LT. 20 .OR. N2 .LT. 20) STOP
        WRITE (6, 600) N1
 600    FORMAT (1X,38HTHE NUMBER OF OBSERVATIONS CODED 1 IS , I6)
        WRITE (6, 601) N2
 601    FORMAT (1X,38HTHE NUMBER OF OBSERVATIONS CODED 2 IS , I6)
C
C     ********************************************************************
C     * STEP 3.  CALCULATE THE MEAN (UR) AND STANDARD DEVIATION         *
C     * (SIGMAR).                                                       *
```

```
C     ****************************************************************
C
      UR = ((2.0 * N1 * N2) / (N1 + N2)) + 1
      WRITE (6, 602) UR
602   FORMAT (1X,12HTHE MEAN IS , F10.3)
      SIGMAR = SQRT((((2.0 * N1 * N2) * ((2.0 * N1 * N2) - N1 - N2)) /
     *    (((N1 + N2) **2) * (N1 + N2 - 1.0)))
      WRITE (6, 603) SIGMAR
603   FORMAT (1X,26HTHE STANDARD DEVIATION IS , F10.3)
C
C     ****************************************************************
C     * STEP 4.  CALCULATE THE TEST STATISTIC (Z).                  *
C     ****************************************************************
C
      Z = (R - UR) / SIGMAR
      WRITE (6, 604) R
604   FORMAT (1X,22HTHE NUMBER OF RUNS IS , I6)
      WRITE (6, 605) Z
605   FORMAT (1X, 31HTHE VALUE OF THE TEST STATISTIC,
     *    4H IS , F10.3)
      STOP
      END
```

D. Example Problems

Example 1   A consumer testing bureau wished to determine whether or not a selected group of shoppers had a preference for either brand A or brand B applesauce. The following sequence of brands was chosen:

$$A \ A \ B \ A \ A \ A \ B \ A \ B \ B$$
$$A \ B \ A \ A \ A \ B \ B \ A \ B \ A$$
$$A \ A \ B \ A \ B \ B \ A \ B \ A \ A$$
$$B \ B \ A \ B \ B \ B \ A \ B \ B \ B$$

Find the Z statistic for the above sequence of symbols.

Input Data

1	1	2	3	4
	0	0	0	0

1
1
2
1
1
1
2
1
2
2
1
2
1
1
1
2
2
1
2
1
1
1
2
1
2
2

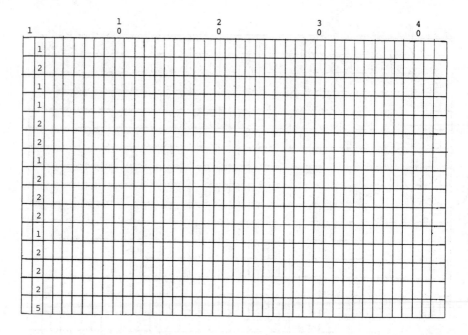

Solution
THE NUMBER OF OBSERVATIONS CODED 1 IS 20
THE NUMBER OF OBSERVATIONS CODED 2 IS 20
THE MEAN IS 21.000
THE STANDARD DEVIATION IS 3.121
THE NUMBER OF RUNS IS 24
THE VALUE OF THE TEST STATISTIC IS 0.961

Example 2   An instructor made up an answer to a 50-item true/false examination and is interested in finding whether or not the sequence of his true and false answers is random. Given the following sequence of letters:

```
T F F F T T F T F F F F
T F T F F F T T T F T F
F F F T T T F T T F T F
F T T T F T F F T F F T
F F
```

Find the Z statistic for the above data.

Input Data

	1									1 0									2 0									3 0									4 0		
1																																										
2																																										
2																																										
2																																										
1																																										
1																																										
2																																										
1																																										
2																																										
2																																										
2																																										
2																																										
1																																										
2																																										
1																																										
2																																										
2																																										
2																																										
1																																										
1																																										
1																																										
2																																										
1																																										
2																																										
2																																										
2																																										
2																																										
1																																										
1																																										

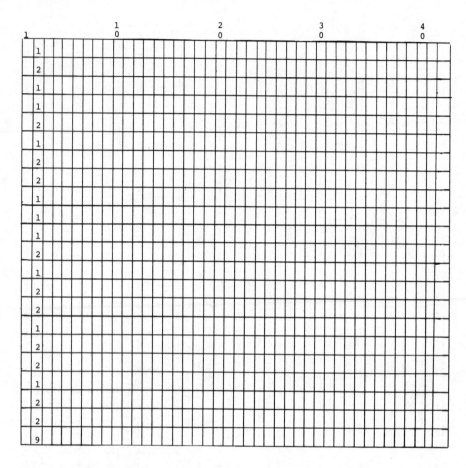

Solution

THE NUMBER OF OBSERVATIONS CODED 1 IS 22
THE NUMBER OF OBSERVATIONS CODED 2 IS 28
THE MEAN IS 25.640
THE STANDARD DEVIATION IS 3.448
THE NUMBER OF RUNS IS 28
THE VALUE OF THE TEST STATISTIC IS 0.684

☆ Mann-Whitney U Test

A. Statement of Problem

The Mann-Whitney U test is a rank test used to test whether two independent samples come from identical parent populations. Let n_1

be the number of variates in sample 1 and n_2 be the number of variates in sample 2. The samples are ranked together by assigning 1 to the lowest variate and 2 to the next lowest variate and so on. The ranks of sample 1 and sample 2 are then computed (ΣR_1 and ΣR_2). The value of the Mann-Whitney U statistic is then computed as:

$$U_1 = n_1 n_2 + \left(\frac{n_1(n_1 + 1)}{2} \right) - \Sigma R_1$$

$$U_2 = n_1 n_2 + \left(\frac{n_2(n_2 + 1)}{2} \right) - \Sigma R_2$$

where the minimal value of U_1 or U_2 is the test statistic U. When both n_1 and $n_2 \geqslant 30$, the Mann-Whitney U statistic is approximately normally distributed with:

Mean $\quad\quad\quad\quad\quad \mu_U = n_1 n_2 / 2$

Standard
deviation $\quad\quad\quad \sigma_U = \sqrt{\dfrac{n_1 n_2 (n_1 + n_2 + 1)}{12}}$

Satisfying a
normal distribution $\quad Z = \dfrac{U - \mu_U}{\sigma_U}$

Write a computer program to find the Mann-Whitney U statistic.

B. Algorithm

1. Determine the number (N) of observations in the analysis. Insure that N does not exceed 1000.
2. Read in all the observations (OBS(J, 1)) and the number of the sample (either 1 or 2) from which the observation was taken (OBS(J, 2)).
3. Sort the observations in ascending order.
4. Determine the sum of the ranks for each sample (SRANK1, SRANK2) and the number of observations in each sample (N1, N2).
5. Determine the mean (UU) and the standard deviation (SIGMAU).
6. Calculate U1 and U2 according to the given formulas.
7. Select the smaller value of U1 and U2 and store this in the variable U.
8. Calculate the Z value, as indicated in the given formula.

C. General Program

```
C
C
C     ***************************************************************
C     * THIS PROGRAM PERFORMS A MANN-WHITNEY U TEST.  IT ASSUMES NO TIES*
C     * IN THE INPUT DATA AND THAT THE SAMPLES ARE NUMBERED 1 AND 2.    *
C     ***************************************************************
C
      INTEGER SRANK1, SRANK2, OBS
      DIMENSION OBS(1000, 2)
C
C     ***************************************************************
C     * STEP 1.  DETERMINE THE TOTAL NUMBER OF OBSERVATIONS IN THE ANAL-*
C     * YSIS (N) AND INSURE THAT IT DOES NOT EXCEED 1000.              *
C     ***************************************************************
C
      READ (5, 500) N
  500 FORMAT (I6)
      IF (N .GT. 1000) STOP

C
C     ***************************************************************
C     * STEP 2.  READ IN ALL THE OBSERVATIONS (OBS(J,1)) AND THE CORRE- *
C     * SPONDING SAMPLE NUMBER (OBS(J,2)) FROM WHICH EACH WAS DRAWN.   *
C     * EACH ROW OF THE ARRAY OBS WILL CONTAIN AN OBSERVATION (IN THE  *
C     * FIRST COLUMN) AND THE SAMPLE NUMBER FROM WHICH IT WAS DRAWN (IN *
C     * THE SECOND COLUMN).  THE SAMPLES MUST BE NUMBERED 1 OR 2.      *
C     ***************************************************************
C
      READ (5, 501) ((OBS(J, L), L = 1, 2), J = 1, N)
  501 FORMAT (2I6)
C
C     ***************************************************************
C     * STEP 3.  SORT THE OBSERVATIONS IN ASCENDING ORDER.  THIS IN-   *
C     * VOLVES COMPARING EACH OBSERVATION EXCEPT THE LAST WITH EVERY   *
C     * OTHER OBSERVATION IN THE ARRAY WITH A GREATER ARRAY ELEMENT    *
C     * NUMBER.  IF THE FIRST ELEMENT IS LESS THAN THE SECOND ELEMENT  *
C     * THE FIRST ELEMENT IS THEN COMPARED TO THE THIRD, FOURTH, ETC.  *
C     * ELEMENT.  IF THE FIRST ELEMENT IS GREATER THAN THE SECOND ELE- *
C     * MENT THE ARRAY ELEMENTS ARE INTERCHANGED (BOTH THE OBSERVATION *
C     * AND THE CORRESPONDING SAMPLE NUMBER).  THE FIRST ELEMENT OF THE *
C     * ARRAY IS THEN COMPARED WITH THE THIRD, FOURTH, ETC. ELEMENT.   *
C     * THE PROCESS CONTINUES BY COMPARING THE SECOND THROUGH THE SECOND*
C     * LAST OBSERVATION TO ALL OTHER OBSERVATIONS WITH GREATER ARRAY  *
C     * ELEMENT NUMBERS AND INTERCHANGING THE POSITIONS IF THE FIRST IS *
C     * GREATER THAN THE SECOND.                                       *
C     ***************************************************************
C
C
C
C     ***************************************************************
C     * SET UP THE OUTER LOOP TO BE CYCLED THROUGH N MINUS 1 TIMES USING*
C     * J AS AN INDEX.                                                 *
C     ***************************************************************
C
      NLESS1 = N - 1
      DO 10 J = 1, NLESS1
C
C     ***************************************************************
C     * SET UP THE INNER LOOP TO BE CYCLED THROUGH N MINUS J TIMES USING*
C     * I AS THE INDEX.  THE INITIAL VALUE OF THE INDEX IS THE VALUE OF *
C     * J PLUS 1.                                                      *
C     ***************************************************************
C
      M = J + 1
      DO 10 I = M, N
```

232

```
C
C    *****************************************************************
C    * IF THE J TH OBSERVATION IS LESS THAN THE I TH OBSERVATION DO NOT*
C    * INTERCHANGE THE OBSERVATIONS.                                  *
C    *****************************************************************
C
      IF (OBS(J, 1) .LT. OBS(I, 1)) GO TO 10
C
C    *****************************************************************
C    * IF THE J TH OBSERVATION IS EQUAL TO THE I TH OBSERVATION THERE *
C    * IS A TIE IN THE INPUT DATA.  SINCE IT IS ASSUMED THAT THERE ARE *
C    * NO TIES THE PROGRAM SHOULD BE TERMINATED.                      *
C    *****************************************************************
C
      IF (OBS(J,1) .EQ. OBS(I, 1)) GO TO 70
C
C    *****************************************************************
C    * AT THIS POINT IT HAS BEEN DETERMINED THAT THE J TH OBSERVATION *
C    * IS GREATER THAN THE I TH OBSERVATION.  INTERCHANGE THE TWO     *
C    * OBSERVATIONS AND THEIR CORRESPONDING SAMPLE NUMBERS.           *
C    *****************************************************************
C
      TEMP1 = OBS(J, 1)
      TEMP2 = OBS(J, 2)
      OBS(J, 1) = OBS(I, 1)
      OBS(J, 2) = OBS(I, 2)
      OBS(I, 1) = TEMP1
      OBS(I, 2) = TEMP2
   10 CONTINUE
C
C    *****************************************************************
C    * STEP 4.  ACCUMULATE THE SUM OF THE RANKS (SRANK1, SRANK2) AND  *
C    * THE SAMPLE SIZES (N1, N2) FOR BOTH SAMPLES.                    *
C    *****************************************************************
C
C
C    *****************************************************************
C    * INITIALIZE THE ACCUMULATORS OF THE SUMS OF THE RANKS AND SAMPLE *
C    * SIZES.                                                         *
C    *****************************************************************
C
      SRANK1 = 0
      SRANK2 = 0
      N1 = 0
      N2 = 0
      DO 20 J = 1, N
C
C    *****************************************************************
C    * SKIP OVER SAMPLES NOT NUMBERED 1.                             *
C    *****************************************************************
C
      IF (OBS(J, 2) .NE. 1) GO TO 18
C
C    *****************************************************************
C    * ACCUMULATE THE SUM OF THE RANKS FOR SAMPLE 1.  SINCE THE OBSER- *
C    * VATIONS WERE SORTED INTO ASCENDING ORDER THE RANK OF THE J TH  *
C    * OBSERVATION IS J.                                             *
C    *****************************************************************
C
      SRANK1 = SRANK1 + J
C
C    *****************************************************************
C    * ACCUMULATE THE NUMBER OF OBSERVATIONS IN SAMPLE NUMBER 1.      *
```

233

```
C
C     *********************************************************************
C
      N1 = N1 + 1
      GO TO 20
C
C
C     *********************************************************************
C     * IF THE SAMPLE NUMBER IS NEITHER A 1 NOR A 2 THE INPUT DATA IS IN*
C     * ERROR.                                                          *
C     *********************************************************************
C
   18 IF (OBS(J, 2) .NE. 2) GO TO 60
C
C
C     *********************************************************************
C     * ACCUMULATE THE SUM OF THE RANKS FOR SAMPLE 2.  SINCE THE OBSER- *
C     * VATIONS WERE SORTED INTO ASCENDING ORDER THE RANK OF THE J TH   *
C     * OBSERVATION IS J.                                               *
C     *********************************************************************
C
      SRANK2 = SRANK2 + J
C
C
C     *********************************************************************
C     * ACCUMULATE THE NUMBER OF OBSERVATIONS IN SAMPLE NUMBER 2.       *
C     *********************************************************************
C
      N2 = N2 + 1
   20 CONTINUE
      WRITE (6, 601) SRANK1
  601 FORMAT (1X,35HTHE SUM OF THE RANKS IN SAMPLE 1 = , I6)
      WRITE (6, 602) SRANK2
  602 FORMAT (1X,35HTHE SUM OF THE RANKS IN SAMPLE 2 = , I6)
C
C     *********************************************************************
C     * STEP 5.  CALCULATE THE MEAN (UU) AND THE STANDARD DEVIATION     *
C     * (SIGMAU).                                                       *
C     *********************************************************************
C
      UU = (N1 * N2) / 2.0
      WRITE (6, 603) UU
  603 FORMAT (1X,12HTHE MEAN IS , F10.3)
      SIGMAU = SQRT(((N1 * N2) * (N1 + N2 + 1)) / 12.0)
      WRITE (6, 604) SIGMAU
  604 FORMAT (1X,26HTHE STANDARD DEVIATION IS , F10.3)
C
C     *********************************************************************
C     * STEP 6.  CALCULATE U1 AND U2.                                   *
C     *********************************************************************
C
      U1 = (N1 * N2) + ((N1 * (N1 + 1)) / 2.0) - SRANK1
      WRITE (6, 605) U1
  605 FORMAT (1X,5HU1 = , F10.3)
      U2 = (N1 * N2) + ((N2 * (N2 + 1)) / 2.0) - SRANK2
      WRITE (6, 606) U2
  606 FORMAT (1X,5HU2 = , F10.3)
C
C     *********************************************************************
C     * STEP 7.  SET U EQUAL TO THE SMALLER OF U1 AND U2.               *
C     *********************************************************************
C
      U = U1
      IF (U2 .LT. U1) U = U2
      WRITE (6, 607) U
  607 FORMAT (1X,4HU = , F10.3)
```

```
C
C
C      ******************************************************************
C      * STEP 9.  CALCULATE Z.                                         *
C      ******************************************************************
C
       Z = (U - UU) / SIGMAU
       WRITE (6, 608) Z
608    FORMAT (1X,18HTHE VALUE OF Z IS , F10.3)
       STOP
60     WRITE (6, 609)
609    FORMAT (1X,32HSAMPLES ARE NOT NUMBERED 1 AND 2)
       STOP
70     WRITE (6, 610)
610    FORMAT (1X,56HTIE ENCOUNTERED.  DATA DOES NOT MEET PROGRAM RESTRICT
      *IONS)
       STOP
       END
70     PRINT, 'TIE ENCOUNTERED.  DATA DOES NOT MEET PROGRAM RESTRICTIONS'
       STOP
       END
```

D. Example Problems

Twelve inner-city students were matched with an equal number of suburban students on a college aptitude examination. The inner-city students (X_i) and the suburban students (Y_i) scores are as follows:

Example 1

X_i		Y_i	
65	59	83	75
45	63	79	86
71	55	68	85
32	39	92	57
51	80	81	72
82	64	73	70

Test the null hypothesis of whether inner-city youths perform as well as suburban youths on the college aptitude examination ($\alpha = 0.01$).

Input Data

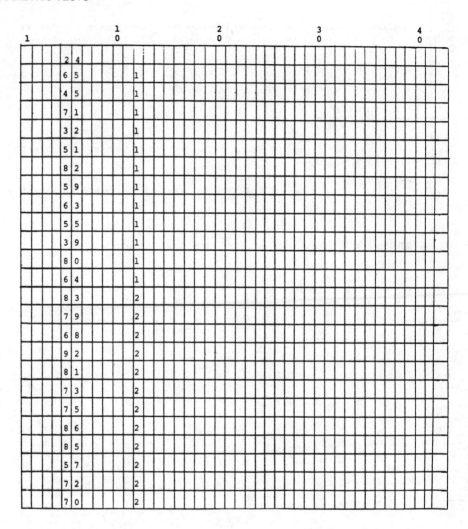

Solution   THE SUM OF THE RANKS IN SAMPLE 1 = 100

THE SUM OF THE RANKS IN SAMPLE 2 = 200

THE MEAN IS 72.000

THE STANDARD DEVIATION IS 17.321

U1 = 122.000

U2 = 22.000

U = 22.000

THE VALUE OF Z IS −2.887

The number of cigarette packages smoked per year were reported   *Example 2*
by men and women as indicated below:

Men	Women
260	650
250	420
200	100
350	570
530	1500
780	1800
390	900
1120	160

Test the null hypothesis of whether the men smoke an equivalent
number of cigarette packages as the women ($\alpha = 0.05$).

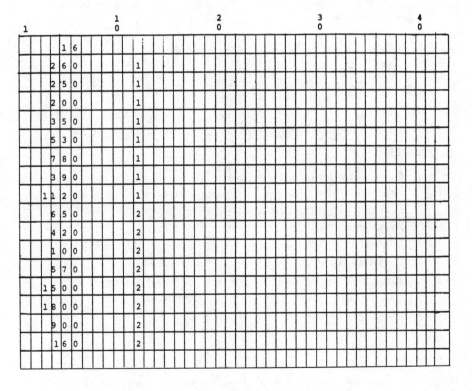

Input Data

Solution   THE SUM OF THE RANKS IN SAMPLE 1 = 60
THE SUM OF THE RANKS IN SAMPLE 2 = 76
THE MEAN IS 32.000
THE STANDARD DEVIATION IS 9.522
U1 = 40.000
U2 = 24.000
U = 24.000
THE VALUE OF Z IS −0.840

☆ Kruskal-Wallis Test

A. Statement of Problem

The Kruskal-Wallis, or H statistic, is a ranked analysis of variance test for two or more independent groups. Suppose that k samples of sizes, $n_1, n_2, \ldots, n_k$ are drawn at random from k identically distributed populations. If the $n_1 + n_2 + \ldots + n_k = n$ observations are ranked in one array according to magnitude and the rank sums ΣR_k for each sample are determined, then the Kruskal-Wallis (H) statistic can be determined from the following equation:

$$H = \frac{12}{n(n + 1)} \left(\Sigma \frac{(\Sigma R_k)^2}{n_k} \right) - 3(n + 1)$$

Write a computer program to find the Kruskal-Wallis H statistic.

B. Algorithm

1. Determine the total number of observations (N) in the analysis and insure it is not greater than 1000. Determine the number of samples (K).
2. Read in all the observations (OBS(J, 1)) and the number of the sample from which each was drawn (OBS(J, 2)).
3. Sort the observations in ascending order.
4. Accumulate the sum of the squares of the ranks divided by the sample sizes (SUMR2N) by performing the following for each sample:
   a. Sum the ranks of the observations (SUMRNK) in the sample.
   b. Count the number of observations in the sample (NI).
   c. Accumulate the sum of the squares of the sum of the ranks divided by the sample size.

5. Calculate the *H* statistic by evaluating the given formula.

C. General Program

```
C     ***********************************************************
C     * THIS PROGRAM PERFORMS A KRUSKAL-WALLIS ONE WAY ANALYSIS OF VAR- *
C     * IANCE BY RANK.  IT ASSUMES NO TIES IN THE RANKS OF THE INPUT   *
C     * DATA.                                                           *
C     ***********************************************************
C
      INTEGER SUMRNK, OBS
      DIMENSION OBS(1000, 2)
C
C     ***********************************************************
C     * STEP 1.  DETERMINE THE NUMBER OF OBSERVATIONS (N) AND THE NUMBER *
C     * OF SAMPLES (K).  INSURE THAT THE NUMBER OF OBSERVATIONS IS NOT   *
C     * GREATER THAN 1000.                                              *
C     ***********************************************************
C
      READ (5, 500) N, K
  500 FORMAT (2I6)
      IF (N.GT.1000) STOP
C
C     ***********************************************************
C     * STEP 2.  READ IN ALL THE OBSERVATIONS (OBS(J,1)) AND THE CORRE- *
C     * SPONDING SAMPLE NUMBER (OBS(J,2)) FROM WHICH EACH WAS DRAWN.    *
C     * EACH ROW OF THE ARRAY OBS WILL CONTAIN AN OBSERVATION (IN THE   *
C     * FIRST COLUMN) AND THE SAMPLE NUMBER FROM WHICH IT WAS DRAWN (IN *
C     * THE SECOND COLUMN).  THE SAMPLES MUST BE NUMBERED SEQUENTIALLY  *
C     * AS 1, 2, . . . TO A MAXIMUM OF 50.                             *
C     ***********************************************************
C
      READ (5, 501) ((OBS(J,L),L = 1, 2), J = 1, N)
  501 FORMAT (2I6)
C
C     ***********************************************************
C     * STEP 3.  SORT THE OBSERVATIONS IN ASCENDING ORDER.  THIS IN-   *
C     * VOLVES COMPARING EACH OBSERVATION EXCEPT THE LAST WITH EVERY   *
C     * OTHER OBSERVATION IN THE ARRAY WITH A GREATER ARRAY ELEMENT    *
C     * NUMBER.  IF THE FIRST ELEMENT IS LESS THAN THE SECOND ELEMENT  *
C     * THE FIRST ELEMENT IS THEN COMPARED TO THE THIRD, FOURTH, ETC.  *
C     * ELEMENT.  IF THE FIRST ELEMENT IS GREATER THAN THE SECOND ELE- *
C     * MENT THE ARRAY ELEMENTS ARE INTERCHANGED (BOTH THE OBSERVATION *
C     * AND THE CORRESPONDING SAMPLE NUMBER).  THE FIRST ELEMENT OF THE *
C     * ARRAY IS THEN COMPARED WITH THE THIRD, FOURTH, ETC. ELEMENT.   *
C     * THE PROCESS CONTINUES BY COMPARING THE SECOND THROUGH THE SECOND*
C     * LAST OBSERVATION TO ALL OTHER OBSERVATIONS WITH GREATER ARRAY  *
C     * ELEMENT NUMBERS AND INTERCHANGING THE POSITIONS IF THE FIRST IS *
C     * GREATER THAN THE SECOND.                                        *
C     ***********************************************************
C
C
C     ***********************************************************
C     * SET UP THE OUTER LOOP TO BE CYCLED THROUGH N MINUS 1 TIMES USING*
C     * J AS AN INDEX.                                                  *
C     ***********************************************************
C
      NLESS1 = N - 1
      DO 10 J = 1, NLESS1
C
C     ***********************************************************
C     * SET UP THE INNER LOOP TO BE CYCLED THROUGH N MINUS J TIMES USING*
C     * I AS THE INDEX.  THE INITIAL VALUE OF THE INDEX IS THE VALUE OF *
C     * J PLUS 1.                                                       *
```

```
C     *******************************************************************
C
      M = J + 1
      DO 10 I = M, N
C
C     *******************************************************************
C     * IF THE J TH OBSERVATION IS LESS THAN THE I TH OBSERVATION DO NOT*
C     * INTERCHANGE THE OBSERVATIONS.                                   *
C     *******************************************************************
C
      IF (OBS(J,1) .LT. OBS(I,1)) GO TO 10
C
C     *******************************************************************
C     * IF THE J TH OBSERVATION IS EQUAL TO THE I TH OBSERVATION THERE  *
C     * IS A TIE IN THE INPUT DATA.  SINCE IT IS ASSUMED THAT THERE ARE *
C     * NO TIES THE PROGRAM SHOULD BE TERMINATED.                       *
C     *******************************************************************
C
      IF (OBS(J,1) .EQ. OBS(I,1)) GO TO 30
C
C     *******************************************************************
C     * AT THIS POINT IT HAS BEEN DETERMINED THAT THE J TH OBSERVATION  *
C     * IS GREATER THAN THE I TH OBSERVATION.  INTERCHANGE THE TWO      *
C     * OBSERVATIONS AND THEIR CORRESPONDING SAMPLE NUMBERS.            *
C     *******************************************************************
C
      TEMP1 = OBS(J,1)
      TEMP2 = OBS(J,2)
      OBS(J,1) = OBS(I,1)
      OBS(J,2) = OBS(I,2)
      OBS(I,1) = TEMP1
      OBS(I,2) = TEMP2
   10 CONTINUE
      WRITE (6, 600)
  600 FORMAT (1X, 40HSAMPLE NUMBER    OBSERVATION          RANK)
C
C     *******************************************************************
C     * STEP 4.   ACCUMULATE THE SUM OF THE SQUARES OF THE SUM OF THE   *
C     * RANKS DIVIDED BY THE SAMPLE SIZES (SUMR2N).                     *
C     *******************************************************************
C
C
C
C     *******************************************************************
C     * INITIALIZE     THE ACCUMULATOR (SUMR2N) USED TO SUM THE SQUARES *
C     * OF THE RANKS (SUMRNK) DIVIDED BY THE SAMPLE SIZE (NI).          *
C     *******************************************************************
C
      SUMR2N = 0.0
C
C     *******************************************************************
C     * SET UP AN OUTER LOOP TO BE CYCLED THROUGH K TIMES USING I AS THE*
C     * INDEX.  THIS VARIABLE, I, WILL BE USED TO GENERATE THE SAMPLE   *
C     * NUMBERS 1 AND 2.                                                *
C     *******************************************************************
C
      DO 40 I = 1, K
C
C     *******************************************************************
C     * INITIALIZE THE ACCUMULATOR (SUMRNK) USED TO ACCUMULATE THE SUMS *
C     * OF THE RANKS FOR EACH SAMPLE.                                   *
C     *******************************************************************
C
      SUMRNK = 0
      NI = 0
```

```
C
C     ************************************************************
C     * SET UP AN INNER LOOP TO BE CYCLED THROUGH N TIMES.      *
C     ************************************************************
C
      DO 20 J = 1, N
C
C     ************************************************************
C     * SKIP OVER THE J TH OBSERVATION IF IT IS NOT FROM THE I TH  *
C     * SAMPLE.                                                  *
C     ************************************************************
C
      IF (OBS(J,2) .NE. I) GO TO 20
C
C     ************************************************************
C     * STEP 4(A).  SUM THE RANKS FOR THE I TH SAMPLE.  SINCE ALL THE *
C     * OBSERVATIONS WERE SORTED IN ASCENDING ORDER, THE RANK OF THE  *
C     * J TH OBSERVATION IS J.                                   *
C     ************************************************************
C
      SUMRNK = SUMRNK + J
C
C     ************************************************************
C     * STEP 4(B).  ACCUMULATE THE NUMBER OF OBSERVATIONS (NI) IN THIS *
C     * SAMPLE.                                                  *
C     ************************************************************
C
      NI = NI + 1
      SAVENI = NI
      WRITE (6, 601) I, OBS(J,1), J
  601 FORMAT (1X, 6X, I6, 4X, I9, 7X, I6)
   20 CONTINUE
C
C     ************************************************************
C     * STEP 4(C).  ACCUMULATE THE SUM OF THE SQUARES (SUMR2N) OF THE *
C     * SUM OF THE RANKS (SUMRNK) DIVIDED BY THE SAMPLE SIZE (NI). *
C     ************************************************************
C
      SUMR2N = SUMR2N + ((SUMRNK * SUMRNK) / SAVENI)
      WPITE (6, 602) I, SUMRNK
  602 FORMAT (1X, 31HTHE SUM OF THE RANKS IN COLUMN , I6, 4H IS , I6)
   40 CONTINUE
C
C     ************************************************************
C     * STEP 5.  CALCULATE THE H STATISTIC AS INDICATED.        *
C     ************************************************************
C
      H = (12.0 / (N * (N + 1))) * SUMR2N - (3 * (N + 1))
      WRITE (6, 603) H
  603 FORMAT (1X, 4HH = , F10.3)
      STOP
   30 WRITE (6, 604)
  604 FORMAT (1X, 57HDATA DOES NOT MEET PROGRAM RESTRICTIONS.  TIE ENCOU
     $NTERED)

      ST P
      END
```

D. Example Problems

Example 1   Four varieties of corn yield the following number of bushels of grain:

A	B	C	D
85	92	55	61
77	96	59	64
65	88	71	57
58	79	73	53
63	84	49	48
	101		
	91		

Determine whether or not the four varieties of corn yield equivalent amounts of grain ($\alpha = 0.01$).

Input Data

	1	10	20	30	40

Observation	Sample
2 2	4
8 5	1
7 7	1
6 5	1
5 8	1
6 3	1
9 2	2
9 6	2
8 8	2
7 9	2
8 4	2
1 0 1	2
9 1	2
5 5	3
5 9	3
7 1	3
7 3	3
4 9	3
6 1	4
6 4	4
5 7	4
5 3	4
4 8	4

Solution

SAMPLE NUMBER	OBSERVATION	RANK
1	58	6
1	63	9
1	65	11
1	77	14
1	85	17

THE SUM OF THE RANKS IN COLUMN 1 IS 57

2	79	15
2	84	16
2	88	18
2	91	19
2	92	20
2	96	21
2	101	22

THE SUM OF THE RANKS IN COLUMN 2 IS 131

3	49	2
3	55	4
3	59	7
3	71	12
3	73	13

THE SUM OF THE RANKS IN COLUMN 3 IS 38

4	48	1
4	53	3
4	57	5
4	61	8
4	64	10

THE SUM OF THE RANKS IN COLUMN 4 IS 27

$H = 14.857$

Example 2   Three competing automobile agencies reported their total monthly car sales for a period of six months. The results are shown below:

A	B	C
50	73	55
48	71	46
53	68	43
47	59	41
51	72	44
35	70	38

Determine whether or not the sales of the three companies were equivalent over the indicated time period ($\alpha = 0.05$).

Input Data

Observation	Sample
1 8	3
5 0	1
4 8	1
5 3	1
4 7	1
5 1	1
3 5	1
7 3	2
7 1	2
6 8	2
5 9	2
7 2	2
7 0	2
5 5	3
4 6	3
4 3	3
4 1	3
4 4	3
3 8	3

Solution

SAMPLE NUMBER	OBSERVATION	RANK
1	35	1
1	47	7
1	48	8
1	50	9
1	51	10
1	53	11

THE SUM OF THE RANKS IN COLUMN 1 IS 46

2	59	13
2	68	14
2	70	15
2	71	16
2	72	17
2	73	18

THE SUM OF THE RANKS IN COLUMN 2 IS 93

3	38	2
3	41	3
3	43	4
3	44	5
3	46	6
3	55	12

THE SUM OF THE RANKS IN COLUMN 3 IS 32

H = 11.941

☆ Friedman Test

A. Statement of Problem

Friedman's two-way analysis of variance test is a rank test for k correlated samples. Write a computer program to compute the quantity

$$\chi^2 = \frac{12}{NC(C + 1)} \left(\sum_{i=1}^{C} R_i^2 \right) - 3N(C + 1)$$

where C is the number of treatments.

N is the total number of observations.

R_i is the sum of the ranks for column i.

B. Algorithm

1. Determine the number of rows (N) and the number of columns (C). Insure that N is greater than 2. Insure that C is greater than 2 but not greater than 50.

2. Do the following for each row of data:

   a. Read in a row of data (OBS(J, 1)).

   b. Sort the row of observations into ascending order.

   c. Accumulate the sum of the ranks (SUMRJ(J)) by column.

3. Accumulate the sum of the squares (SUMR2) of the sum of the ranks (SUMRJ(J)).

4. Compute the chi square statistic by evaluating the given formula.

C. General Program

```
C
C       *************************************************************
C       * THIS PROGRAM PERFORMS A FRIEDMAN TWO-WAY ANALYSIS OF VARIANCE BY*
C       * RANKS.                                                        *
C       *************************************************************
C
        INTEGER SUMRJ, OBS, COLSUB, SUMR2, C, CLESS1
        DIMENSION SUMRJ(50), OBS(50,2)
C
C       *************************************************************
C       * STEP 1.  DETERMINE THE NUMBER (N) OF ROWS AND THE NUMBER (C) OF *
C       * COLUMNS.  INSURE THAT THE NUMBER OF ROWS IS NOT LESS THAN 2 AND *
C       * THAT THE NUMBER OF COLUMNS IS IN THE RANGE 2 TO 50.            *
C       *************************************************************
C
        READ (5, 500) N,C
  500   FORMAT (2I6)
        IF (N.LT.2) STOP
        IF (C.LT.2 .OR. C.GT.50) STOP
C
C       *************************************************************
C       * INITIALIZE THE ARRAY USED TO ACCUMULATE THE SUM OF THE RANKS  *
C       * (SUMRJ(J)) IN EACH COLUMN.                                    *
C       *************************************************************
C
        DO 10 J = 1, C
  10    SUMRJ(J) = 0
C       *************************************************************
C       * STEP 2.  SET UP A LOOP TO PROCESS ALL ROWS OF DATA.           *
C       *************************************************************
C
        DO 30 I = 1, N
C
C       *************************************************************
C       * STEP 2(A).  READ IN THE ITH ROW OF DATA.                      *
C       *************************************************************
C
        READ (5, 501) (OBS(J,1), J = 1, C)
  501   FORMAT (I6)
C
C       *************************************************************
C       * STORE THE ORIGINAL COLUMN NUMBER OF OF EACH OBSERVATION.      *
C       *************************************************************
C
        DO 15 J = 1, C
  15    OBS(J,2) = J
C
C       *************************************************************
C       * STEP 2(B).  SORT THE ROW OF OBSERVATIONS INTO ASCENDING ORDER. *
C       *************************************************************
C
        CLESS1 = C - 1
        DO 20 J = 1, CLESS1
        JPLUS1 = J + 1
        DO 20 L = JPLUS1, C
```

```
          IF (OBS(J,1) .LT. OBS(L,1)) GO TO 20
          IF (OBS(J,1) .EQ. OBS(L,1)) GO TO 45
          TEMP1 = OBS(J,1)
          TEMP2 = OBS(J,2)
          OBS(J,1) = OBS(L,1)
          OBS(J,2) = OBS(L,2)
          OBS(L,1) = TEMP1
          ORS(L,2) = TEMP2
      20  CONTINUE
C
C
C     ***********************************************************************
C     * STEP 2(C).  ACCUMULATE THE SUM OF THE RANKS (SUMRJ(J)) BY           *
C     * COLUMN.                                                             *
C     ***********************************************************************
C
C
C
C     ***********************************************************************
C     * SET UP A LOOP TO PROCESS ALL COLUMNS OF THIS ROW OF SORTED DATA.*
C     ***********************************************************************
C
          DO 25 J = 1, C
C
C     ***********************************************************************
C     * SET COLSUB EQUAL TO THE ORIGINAL COLUMN NUMBER OF THIS OBSERVA- *
C     * TION.                                                           *
C     ***********************************************************************
C
          COLSUB = OBS(J,2)
C
C     ***********************************************************************
C     * ACCUMULATE THE SUM OF THE RANKS (SUMRJ(J)) BY COLUMN.               *
C     ***********************************************************************
C
      25  SUMRJ(COLSUB) = SUMRJ(COLSUB) + J
      30  CONTINUE
C
C     ***********************************************************************
C     * STEP 3.  ACCUMULATE THE SUM OF THE SQUARES (SUMR2) OF THE SUM OF*
C     * THE RANKS (SUMRJ(J)) OF EACH COLUMN.                            *
C     ***********************************************************************
C
C
C     ***********************************************************************
C     * INITIALIZE THE ACCUMULATOR (SUMR2) OF THE SUM OF THE SQUARES OF *
C     * THE SUMS OF THE RANKS.                                          *
C     ***********************************************************************
C
          SUMR2 = 0
C
C     ***********************************************************************
C     * SET UP THE LOOP TO PROCESS EACH COLUMN OF RANK TOTALS.              *
C     ***********************************************************************
C
          DO 40 J = 1, C
          WRITE (6, 600) J, SUMRJ(J)
      600 FORMAT (1X, 31HTHE SUM OF THE RANKS IN COLUMN , I6, 4H IS , I6)
C
C     ***********************************************************************
C     * SUM THE SQUARES OF THE SUM OF THE RANKS.                            *
C     ***********************************************************************
C
      40  SUMR2 = SUMR2 + (SUMRJ(J) * SUMRJ(J))
C
C     ***********************************************************************
C     * STEP 4.  COMPUTE THE CHI SQUARE STATISTIC (CHISQR) BY EVALUATING*
```

248

```
C     * THE GIVEN FORMULA.                                              *
C     ****************************************************************
C
      CHISQR = ((12.0 / (N * C * (C + 1))) * SUMR2 - (3 * N * (C + 1)))
      WRITE (6, 601) CHISQR
601   FORMAT (1X, 27HTHE VALUE OF CHI SQUARE IS , F10.3)
      STOP
45    WRITE (6, 602)
602   FORMAT (1X, 57HTIE ENCOUNTERED.  DATA DOES NOT MEET PROGRAM RESTRI
     $CTIONS)
      STOP
      END
```

D. Example Problems

Five beauty pageant contestants were evaluated by seven judges on
the basis of 1 (highest) to 10 (lowest) points. The results are indicated
below:

Example 1

Judge	Contestant rating				
	I	II	III	IV	V
1	8	2	3	6	5
2	5	1	2	4	3
3	6	3	1	2	5
4	7	4	3	8	6
5	10	2	3	6	4
6	9	3	1	2	5
7	6	4	2	1	7

Can one accept the null hypothesis of equivalent ranks for the above
data at the $\alpha = 0.05$ level of significance?

Input Data

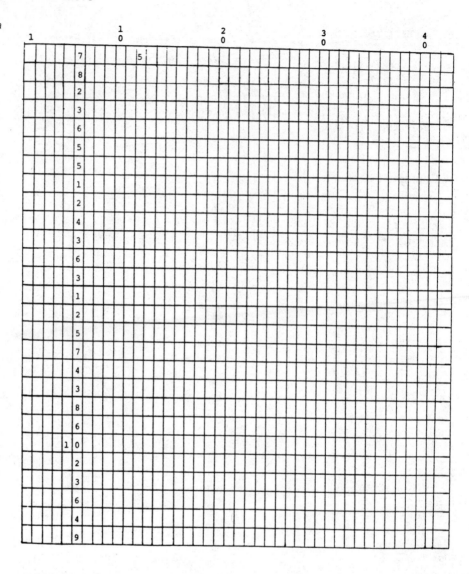

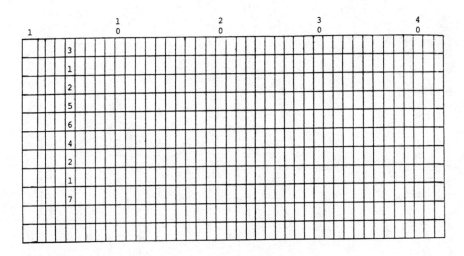

THE SUM OF THE RANKS IN COLUMN 1 IS 33

THE SUM OF THE RANKS IN COLUMN 2 IS 14

THE SUM OF THE RANKS IN COLUMN 3 IS 11

THE SUM OF THE RANKS IN COLUMN 4 IS 22

THE SUM OF THE RANKS IN COLUMN 5 IS 25

THE VALUE OF CHI SQUARE IS 17.714

Solution

Example 2

Five different brands of aspirin were evaluated by a six-member consumer panel as to their effectiveness on the basis of their degree of relief from 1 (excellent) to 10 (no effect). The results are indicated below:

Panel member	Brand				
	A	B	C	D	E
1	8	3	2	5	6
2	4	2	6	3	5
3	6	3	5	4	8
4	3	2	7	6	5
5	7	4	6	9	5
6	5	2	7	6	4

Can we accept the null hypothesis of equivalent ranks at the $\alpha = 0.05$ level of significance?

Input Data

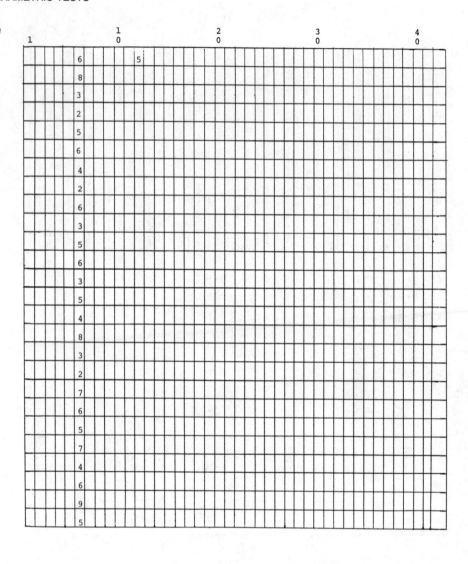

THE SUM OF THE RANKS IN COLUMN 1 IS 21

Solution

THE SUM OF THE RANKS IN COLUMN 2 IS   7

THE SUM OF THE RANKS IN COLUMN 3 IS 22

THE SUM OF THE RANKS IN COLUMN 4 IS 20

THE SUM OF THE RANKS IN COLUMN 5 IS 20

THE VALUE OF CHI SQUARE IS 10.267

☆ Exercises

1. A sporting goods store wished to determine whether or not its customers had a preference for either brand A or brand B tennis balls. The results of one week's purchases are indicated below:

   A B B B A A B B A B B A B B B A A B A
   B B A B B B A A B A B B B A B A B B A B B

   Test the data for randomness by computing the Z statistic ($\alpha = 0.05$).

2. Two adjoining restaurants tabulated their gross sales for a one-week period. Their individual sales totals are listed below:

Restaurant A	370	512	650	380	610	780	1080
Restaurant B	210	402	240	112	430	512	615

   Test whether or not there is a significant difference between the mean sales of the two restaurants ($\alpha = 0.01$).

3. Three groups of rats were administered separate types of hormone drugs and were then required to run a maze. The times for completion of the maze are indicated below:

Group A	12.1	11.5	10.8	13.2	12.8
Group B	8.6	7.3	8.1	7.2	9.1
Group C	10.2	11.3	10.7	12.4	13.6

   Test the null hypothesis of whether the three response times are equivalent ($\alpha = 0.05$).

4. Seven of the top ten USLTA's ranked tennis players rated four types of tennis surfaces from 1 (most preferable) to 8 (least preferable). The results of those ratings are as follows:

Player	Surface			
	1	2	3	4
1	6	3	2	5
2	7	4	1	8
3	5	4	2	6
4	8	7	3	6
5	4	5	2	7
6	5	6	1	8
7	6	5	4	7

Test the above data to determine whether there is a preference among the players for a particular surface ($\alpha = 0.01$).